PCB DESIGN FOR REAL-WORLD EMI CONTROL

THE KLUWER INTERNATIONAL SERIES
IN ENGINEERING AND COMPUTER SCIENCE

PCB DESIGN FOR REAL-WORLD EMI CONTROL

by

Bruce R. Archambeault
IBM Corporation
Research Triangle Park, NC

KLUWER ACADEMIC PUBLISHERS
Boston / Dordrecht / London

Distributors for North, Central and South America:
Kluwer Academic Publishers
101 Philip Drive
Assinippi Park
Norwell, Massachusetts 02061 USA
Telephone (781) 871-6600
Fax (781) 681-9045
E-Mail: kluwer@wkap.com

Distributors for all other countries:
Kluwer Academic Publishers Group
Post Office Box 322
3300 AH Dordrecht, THE NETHERLANDS
Telephone 31 786 576 000
Fax 31 786 576 474
E-Mail: services@wkap.nl

Electronic Services < http://www.wkap.nl>

Library of Congress Cataloging-in-Publication Data

A C.I.P. Catalogue record for this book is available
from the Library of Congress.

Copyright © 2002 by Kluwer Academic Publishers

All rights reserved. No part of this work may be reproduced, stored in a retrieval system, or transmitted in any form or by any means, electronic, mechanical, photocopying, microfilming, recording, or otherwise, without written permission from the Publisher, with the exception of any material supplied specifically for the purpose of being entered and executed on a computer system, for exclusive use by the purchaser of the work.

Permission for books published in Europe: permissions@wkap.nl
Permissions for books published in the United States of America: permissions@wkap.com

Printed on acid-free paper.

Printed in the United States of America

Dedication

This book is dedicated to my wife, best friend, and the person who is the center of my universe, Susan Archambeault.

Contents

1 Introduction to EMI/EMC Design for Printed Circuit Boards　1

   1.1 Introduction to EMI/EMC　1
   1.2 EMI Emissions Sources　4
   1.3 Inductance　5
   1.4 "Ground"　6
   1.5 Shielding　6
   1.6 Summary　7

2 EMC Fundamentals　9

   2.1 Introduction　9
   2.2 Coupling Mechanisms　10
      2.2.1 Electric Field Coupling　10
      2.2.2 Magnetic Field Coupling　12
   2.3 Signal Spectra　14
      2.3.1 Clock Signals' Harmonic Frequencies　14
      2.3.2 Hertz vs. Bits-per-Second　17
      2.3.3 Non-Squarewave Data Signals　18
   2.4 Resonance Effects　19
      2.4.1 Magic and Luck　20
   2.5 Potential Emissions Sources　21
      2.5.1 Shielded Products　21
      2.5.2 Unshielded Products　22
   2.6 Intentional Signal Content　23
   2.7 Summary　23

3 What is Inductance? 25

3.1 Introduction 25
3.2 Electromagnetic Induction 25
3.3 Mutual Inductance 27
3.4 Self-Inductance 29
    3.4.1 Self-Inductance per Unit Length 33
3.5 Partial Inductance 36
3.6 Summary 40

4 The Ground Myth 43

4.1 Where Did The Term "Ground" Originate? 43
4.2 What Do We Mean When We Say "Ground"? 46
    4.2.1 Signal Reference 46
    4.2.2 Power Reference 49
    4.2.3 Chassis Reference 50
    4.2.4 Unshielded Cables 51
    4.2.5 Shielded Cables 54
    4.2.6 Earth Safety Reference 55
4.3 'Ground' is Not a Current Sink 55
4.4 Referencing Strategies 55
    4.4.1 Single-Point Ground-Reference Strategy 56
    4.4.2 Multi-Point Ground-Reference Strategy 56
4.5 Grounding Heatsinks to PC boards 58
    4.5.1 Heatsink "Grounding" Example 61
4.6 PCB Reference Connection to Chassis Reference 64
    4.6.1 I/O Area Connection 64
4.7 Summary 66

5	**Return Current Design**	**69**
	5.1 Introduction	69
	5.2 Split Reference Planes	71
	5.2.1 Stitching Capacitors	72
	5.3 Trace Changing Reference Planes	76
	5.4 Motherboards and Daughter Cards	80
	5.4.1 Connector Pin Assignments	82
	5.5 Summary	83

6	**Controlling EMI Sources – Intentional Signals**	**85**
	6.1 Introduction	85
	6.2 Critical Signals	86
	6.3 Intentional Signals	86
	6.4 Intentional Signals – Loop-Mode	93
	6.5 Controlling Emissions from Intentional Signals – Loop-Mode	95
	6.6 Intentional Signals – Common-mode	96
	6.7 Intentional Signals – Common-mode with Interrupted Return Path	99
	6.7.1 Critical Signal Traces Crossing Splits	99
	6.7.2 Critical Signals Through Vias	100
	6.8 Summary	102

7	**Controlling EMI Sources – Unintentional Signals**	**105**
	7.1 Introduction	105
	7.2 Unintentional Signals	106
	7.3 Unintentional Signals – Common-mode	106

	7.4 Controlling Emissions from Unintentional Signals – Common-mode	108
	7.5 Unintentional Signals – 'Crosstalk' Coupling onto I/O Lines	113
	7.6 Controlling Emissions from Unintentional Signals – 'Crosstalk' Coupling to I/O Lines	115
	7.7 Summary	118
8	**Decoupling Power/Ground Planes**	**121**
	8.1 Introduction	121
	8.2 Background	122
	8.3 Calculating the Source of Decoupling Noise	124
	8.3.1 Decoupling Noise from ASIC/ICs power pins	124
	8.4 Decoupling Capacitor Effectiveness	130
	8.4.1 Test Board Description	131
	8.4.2 Empty Test Board Configuration	134
	8.4.3 Quantity of Distributed (Global) Decoupling Capacitors (.01uf Only)	136
	8.4.4 Quantity of Distributed Decoupling Capacitors (0.01uf and 330 pF)	137
	8.4.5 Selecting the Value of the Decoupling Capacitors	140
	8.4.6 Perfect Decoupling Capacitors	140
	8.4.7 Source Vs Distributed Decoupling	141
	8.4.8 Buried Capacitance Decoupling	144
	8.4.9 Lossy Capacitors	146
	8.5 Summary	148
9	**EMC Filter Design**	**151**
	9.1 Introduction	151
	9.2 Filter Design Concepts	151

	9.3 Filter Configurations	155
	9.3.1 Two-Component Filter Configurations	155
	9.3.2 Reference Connection for Two-Component Filters	157
	9.3.3 Three Component Filter Configurations	161
	9.3.4 Single Component Filter Configurations	163
	9.4 Non-Ideal Components and the Impact on Filters	163
	9.4.1 Non-Ideal Capacitors	164
	9.4.2 Non-Ideal Ferrite Beads	166
	9.4.3 Non-Ideal Zero Ohm Resistors	168
	9.5 Common-Mode Filters	168
	9.6 Summary	169
10	**Using Signal Integrity Tools for EMC Analysis**	**171**
	10.1 Introduction	171
	10.2 Intentional Current Spectrum	172
	10.3 Trace Current for Decoupling Analysis	177
	10.4 Differential Signals Analysis	179
	10.4.1 Internal Differential Signal Lines	181
	10.4.2 External I/O Differential Signal Lines	182
	10.5 Crosstalk Analysis	185
	10.6 Summary	185

11 Printed Circuit Board Layout — 187

11.1 Introduction	187
11.2 PC Board Stack-up	187
11.2.1 Many Layer Boards	188
11.2.2 Six-Layer Boards	191
11.2.3 Four-Layer Boards	192
11.2.4 One and Two-Layer Boards	193
11.3 Component Placement	195
11.4 Isolation	195
11.5 Summary	196

12 Shielding in Enclosures with Apertures — 199

12.1 Introduction	199
12.2 Resonance Mode within Shielded Enclosures	202
12.3 Shielded Enclosures	208
12.3.1 Apertures and Openings	208
12.3.2 Gaskets	210
12.4 Predicting the Shielding Effectiveness of Enclosures with Apertures	213
12.5 Shielding the PC Board Edge	215
12.6 Cable Shields	216
12.7 Summary	218

13 What To Do If a Product Fails in the EMC Lab — 221

13.1 Introduction	221
13.2 Where Does the Signal Come From?	222
13.3 How Does the Signal Get Out of the Shielded Enclosure?	223
13.3.1 Leaks through slots holes and apertures	223

13.3.2 Conducted through the shield on cables and wires	225
13.3.3. Leaks from imperfect mating of shielded cable shields to the enclosure	226
13.4 Coupling Mechanism	227
13.4.1 Case 1 Clock signal leaking from seam	228
13.4.2 Case 2 Clock signal leaking from an unshielded cable	228
13.5 Summary	229

Appendix A Introduction to EMI/EMC Computational Modeling — 231

A.1 Introduction	231
A.2 Why Is EMI/EMC Modeling Important?	232
A.3 EMI/EMC Modeling: State of the Art	233
A.4 Tool Box Approach	234
A.5 Brief Description of EMI Modeling Techniques	235
A.5.1 Finite Difference Time-Domain	235
A.5.2 Method of Moments	237
A.5.3 Finite Element Method	238
A.6 Other Uses for Electromagnetic Modeling	239
A.7 Summary	239

Index — 241

Preface

Proper design of printed circuit boards can make the difference between a product passing emissions requirements during the first cycle or not. Traditional EMC design practices have been simply rule-based, that is, a list of rules-of-thumb are presented to the board designers to implement. When a particular rule-of-thumb is difficult to implement, it is often ignored. After the product is built, it will often fail emission requirements and various time consuming and costly add-ons are then required.

Proper EMC design does not require advanced degrees from universities, nor does it require strenuous mathematics. It *does* require a basic understanding of the underlying principles of the potential causes of EMC emissions. With this basic understanding, circuit board designers can make trade-off decisions during the design phase to ensure optimum EMC design. Consideration of these potential sources will allow the design to pass the emissions requirements the first time in the test laboratory.

A number of other books have been published on EMC. Most are general books on EMC and do not focus on printed circuit board design. This book is intended to help EMC engineers and design engineers understand the potential sources of emissions and how to reduce, control, or eliminate these sources. This book is intended to be a 'hands-on' book, that is, designers should be able to apply the concepts in this book directly to their designs in the real-world.

Chapter 1 provides an introduction into the basic concepts in this book, such as inductance, "ground", and shielding.

Chapter 2 continues the fundamentals of EMC with discussions on various coupling mechanisms, the frequency-domain harmonic content of time-domain signals, resonance phenomena, and potential emissions sources.

Chapter 3 presents a brief introduction to inductance. While the subject of inductance could be an entire book by itself, this chapter presents an overview of self-inductance, mutual-inductance, and partial-inductance. A number of basic formulas are included to find the inductance for simple loop geometries.

Chapter 4 discusses the concept of "ground". There are a number of different meanings for the term "ground", causing this term to be often misused and misunderstood. These different meanings are discussed, and issues with the various purposes of "ground" highlighted.

The next three chapters are the heart of the basic concept of controlling the currents on the printed circuit board. Chapter 5 discusses return current design, and the paths the return current must often take (if proper design practices are not followed). Chapter 6 identifies the potential sources from *intentional signals*, while Chapter 7 identifies the potential sources from *unintentional signals* on the printed circuit board. Both types of signals must be considered in order to properly control the currents on a printed circuit board.

Chapter 8 discusses decoupling power and ground-reference planes on printed circuit boards. This is an area where many myths, misinformation, and unfounded beliefs exist among designers. Chapter 8 also presents the two main sources of EMI 'noise' between the planes, and discusses various decoupling capacitor strategies' effectiveness in reducing this noise.

Chapter 9 presents basic filter concepts for I/O areas on printed circuit boards. The I/O filter is usually the last defense against signals from inside a metal enclosure from being coupled onto the outside cables and causing emissions. Careful design of the I/O filter, and proper connection of the filter capacitor to the appropriate reference, can change an ineffective filter to an effective filter.

Chapter 10 discusses how to use standard signal integrity tools for EMC applications. These tools are usually used by design engineers to ensure functionality of the various high-speed traces on the printed circuit board, and can often help analyze the EMC effectiveness of various design options.

Chapter 11 discusses the printed circuit board stack-up for various board configurations. Careful selection of which routing layer is used for which signals can make a significant difference to the return current path, and ultimately, to the EMC emissions.

Chapter 12 presents an introduction to shielding for printed circuit boards. Resonance effects within the enclosure, the effect of aperture size, and gasket design can all affect the overall shielding and ultimately, the amount of emissions 'leaking' from the enclosure.

Finally, Chapter 13 discusses a strategy to help quickly identify the source of any signals which may cause EMC emissions levels that are over the allowed limit in the EMC test chamber. Random application of copper tape, ferrite beads, and capacitors is not the optimum way to solve emissions problems quickly. The source of the emission should be identified and the 'fix' applied to the source, for the most effective and efficient solution.

The author encourages all readers to not blindly accept rules-of-thumb from so-called experts. It is important to understand the basic concepts for the important issues, and to make the correct decisions for each issue independantly. This does not mean the 'experts' are not correct, because they often *are* correct. This means that you should understand "why" or "why not" to implement a specific design practice, and to use engineering judgement for each circumstance. Design rules-of-thumb are correct for certain circumstances, but ultimately *you* need to make sure they apply to your specific circumstance.

Acknowledgements

The author wishes to gratefully acknowledge the editorial and technical review by Michelle Cook of IBM in Austin, Texas. Her EMC engineering background and editorial skills provided a valuable combination that enhanced the quality of this book. Her repeated reviews of the various revisions of the book helped make this book a success. Recognition, acknowledgement and gratitude is also due to my wife, Susan Archambeault. Her editorial reviews and attention to detail throughout the manuscript enhanced the professional appearance and general readability of the book. Her patience during the writing of this book during the author's mood swings of desperation, exuberance, and despair is greatly appreciated! The author also wishes to acknowledge the technical review from Vince Lisica of IBM in Research Triangle Park, NC. His early review and comments helped fashion the overall content of the book.

Finally, the author wishes to gratefully acknowledge the contributions made by Professor James Drewniak at the University of Missouri-Rolla. His advice helped shape the outline of the book, and his detailed technical review helped ensure that the concepts were clear, correct, and easy to understand. He has been a good friend throughout the writing of this book and made significant contributions in a number of ways to the final manuscript.

Chapter 1

Introduction to EMI/EMC Design for Printed Circuit Boards

1.1 Introduction to EMI/EMC

Electromagnetic Interference and Electromagnetic Compatibility (EMI/EMC) first became a concern in the 1940's and 1950's, mostly as motor noise that was conducted over power lines and into sensitive equipment. During this period, and through the 1960s, EMI/EMC was primarily of interest to the military to ensure electromagnetic compatibility. In a few notable accidents, radar emissions caused inadvertent weapons release, or EMI caused navigation systems failure, so military EMI/EMC was concerned chiefly with electromagnetic compatibility, especially within a weapons system of a airplane or ship.

With the computer proliferation during the 1970s and 1980s, interference from computing devices became a significant problem to broadcast television and radio reception, as well as emergency services radio reception. The U.S. government decided to regulate the amount of electromagnetic emissions from products in this industry. The Federal Communications Commission (FCC) created a set of rules to govern the amount of emissions from any type of computing device and how those emissions were to be measured. Similarly, European and other governments began to limit emissions from computing devices. During this time, EMI/EMC control was limited to computers, peripherals, and computer communications products.

During the 1990's, the concern over EMI/EMC has been found to broaden dramatically; in fact, many countries have instituted import controls requiring that EMI/EMC regulations be met before products can be imported into that country. The compatibility of all devices and equipment must coexist harmoniously in the overall electromagnetic environment. Emissions, susceptibility to emissions from other equipment, susceptibility to electrostatic discharge – all from either radiated or conducted media – are controlled. No longer is this control limited to only computers. Now any product that may potentially radiate (or conduct on cables) EMI, or that could be susceptible to other emissions (or noise conducted into the product), must be carefully tested. Products with no previous need for EMI/EMC control must now comply with the regulations, including dishwashers, video cassette recorders (VCRs), industrial equipment, and most electronic equipment.

While commercial products have come under tighter control for EMI/EMC, the military has not relaxed its EMI/EMC requirements. In fact, because of the higher degree of automation and faster processing speeds, military EMI/EMC control has become a significant part of all military programs.

EMI/EMC design means different things to different people. The standards for commercial applications, such as VCRs, personal computers, and televisions are fairly loose compared to the military/TEMPEST[1] standards, however, they are still difficult to meet. The result of the relaxed nature of these commercial standards is that designers are constantly caught between lowering emissions and susceptibility while meeting cost reduction goals. The trade-offs between EMI/EMC design features are clear, but whether the absolute need for one or another individual EMI/EMC component is required is not so clear. Traditionally, EMI/EMC engineers have used experience, as well as equations and graphs from handbooks, frequently taken out of context, to help during the product design phase. Very little high quality EMI/EMC engineering-level training is available at universities, or at any institutions, and most engineers working in this area find these traditional methods somewhat inadequate.

Military, space, and other government applications must control the emissions of electronics for security, weapon systems functionality, or proper communications, most often to a level far below the commercial emissions/susceptibility level. This increased

---

[1] TEMPEST is the U.S. government code name for the project which controls data related RF emissions from equipment processing classified information.

control requires additional EMI/EMC design features, and greater expense, just when these applications are being forced to reduce costs.

EMI/EMC problems are caused by changes in current with respect to time on conductors within the equipment, known as *di/dt* noise. This current change causes electromagnetic emissions. Alternatively, external electromagnetic energy can induce *di/dt* noise in circuits, causing false logic switching and improper operation of devices. Most high speed fast rise time signals cause EMI/EMC problems. These problems are magnified through the wires and cables attached to the product, creating more efficient antennas at lower frequencies. The typical solution is to use metal shielding, to filter all data/power lines, and to provide significant on-board filtering of signal lines and power planes. The real question is "How much is enough?" and "How much is too much?".

While many EMC engineers have long advocated close attention to EMC design early in the product design cycle, this 'close attention' often was simply a long list of "do's" and "don'ts" in the form of EMC design rules. These rules were most often based on past experience on some specific product or product family, and often these rules contradicted each other. The absence of true, root cause analysis yielded no real understanding of *why* a certain rule was required, or, even more importantly, when a particular rule could not easily be followed (due to some design constraint), then what *alternatives* should be investigated. The designers were simply told "do this or else".

This resulted in a typical design process as follows:
1. EMC engineer provides product designers with list of EMC design rules.
2. Product designers are unable, or unwilling, to implement all the EMC design rules.
3. The product is designed/built including only convenient EMC design rules.
4. Once prototype product is tested in the EMC test chamber, it fails.
5. EMC engineer and design engineer spend 2 weeks to 2 months adding capacitors, ferrite beads, gasket, and finger stock until problem is solved.

6. Product design is updated with EMC 'fixes', and general manufacturing begins.

This design process resulted in delayed first customer shipments of products as well as increased product cost, since the EMC parts were not designed into the product initially. Clearly, this is not a desirable design process, but it is very common!

The goal of this book is to try to remove the magic from the EMC design process. Many have referred to EMC design as black magic, voodoo, simply wild guesses, and worse. EMC design is complex. There are many interconnected events that occur, and most are hard to predict, especially when taken together. If each potential EMC emissions source is taken individually, and proper design approaches can be applied to address each potential source, however, the designer will not get confused while looking at the overall product.

The authors believe that the goal of every design project should be to pass the EMC requirements the *first time* in the EMC test chamber. If a particular IC has more EMI noise than expected and causes a failure, this should be the exception, and not the rule, and the problem should be easily corrected.

Proper EMC design is not just a list of rules. It does require a thought process that considers the potential sources and addresses each in turn. Naturally, the primary goal is to insure the product functions as intended. If, however, during the design phase, these EMC considerations are included and matched to the functionality requirements, then designers can achieve success in both areas with the first design pass.

1.2 EMI Emissions Sources

The most efficient way to reduce EMI emissions is to control the contributing signals at their source. So, where do these signals originate? There are a number of places where these signals can originate, but the majority of the signals start as fast switching currents in ICs.

Nearly all EMI emissions come from common-mode currents that exist someplace within the product. All of these common mode currents come from some intentional current, that is, some current in an IC that is needed for the functionality of the product. If the intentional signal is controlled to only include the harmonics that are necessary for the proper operation of the product, then the chance of a high-frequency harmonic causing unwanted emissions is greatly reduced.

The source of these common mode currents is most likely the intentional signal return current path. While printed circuit board (PCB) layout engineers will take great care to route a trace from source to destination (driver to receiver), very little attention, if any, is given to the return current path. When clock speeds were less than 10 MHz or so, the return current path was not particularly important. Today, on-board clock speeds of 200 – 400 MHz, and data bus speeds above 1 GHz are common, therefore, the signal trace must be considered a microwave transmission line. The return current path at high frequencies is critical, both to the EMC performance and the functionality performance of the signal trace.

1.3 Inductance

One of the most commonly misunderstood concepts is the concept of inductance. Beginning engineering students learn about inductance as specific components in inductors and transformers, but seldom consider the inductance of the current paths in our ground-reference planes, traces, etc. Inductance requires a current flowing through a loop. Sometimes the total path is not clear, and total path may even be partially radiated, so the concept of partial inductance is also important. Partial inductance can be combined to create the total loop inductance. However, if the partial inductance of a portion of the path is considered and reduced, then the total inductance of the path is also reduced.

The high speed signal rates used today make the consideration of loop inductance and partial inductance more important than ever before. Even a perfect superconductor has inductance. The impedance of this inductance provides a voltage potential when a current flows through it. This voltage potential causes noise across ground-reference planes, signal amplitude drop, and adds to EMI emissions.

1.4 "Ground"

If inductance is one of the most commonly misunderstood concepts, "ground" is the MOST misunderstood. When the term "ground" is used, designers mean a variety of different things. It might mean the safety earth reference for the 50/60 Hz AC power. It might mean the signal reference for a high speed trace. It might mean the power return on a PCB, or, it could mean the chassis reference for a shielded chassis. It might even mean the actual earth ground, as in the ground plane at an Open Area Test Site (OATS), where EMI emissions are often measured.

It is clear that it is impossible for all these different uses of the word "ground" to be at the same electric potential, but that is really the intended definition of "ground". "Ground" is the point of zero potential. Actually, "ground" or zero potential only exists at infinity, and so, unless some really long leads are used, true zero ground potential will not exist in our products.

A much more clear and accurate way to describe the various uses of "ground" would be to use the terms: earth-ground, ground-reference, power-reference, chassis-reference, etc. Then both the listener and the speaker are more likely to clearly understand each other.

1.5 Shielding

Another area of confusion in EMC is the concept of shielding. Classical shielding theory shows the effect of a plane wave hitting a shield with an aperture. With the density of our products, designers seldom have a plane wave hitting our apertures in a shield, so this classical approach can be misleading.

Within typical products, the source of the energy is closely coupled to the enclosure shield and apertures. This close capacitive, inductive, or electromagnetic coupling induces currents on the shield. The spacing of the shield to the source is important and can cause the coupling to change dramatically. Any currents induced in the shield can find apertures and transfer energy outside for the enclosure.

1.6 Summary

The most efficient way to approach proper EMC design for PCBs is to consider the various sources and control the signals at their sources. Further chapters will provide much more detail about the different sources and how to control each of the individual sources. EMC must be considered early during the design phase, and indeed, throughout the entire design phase.

It is important not to treat EMC design as simply following a cookbook. During the design, there must be a number of engineering trade-offs. If the designer understands the goals, and understands where the sources of EMI emissions occur and how to control them, then the designer will be successful in making the correct design trade-off. If the EMC design process is to just follow a set of rules, then when these rules become difficult or impossible to implement, they will be ignored, the product will likely fail, and the 'typical' design process explained earlier will repeat itself.

In order to completely understand the thought processes associated with the sources, etc, some basic concepts must also be clearly understood, such as "ground", inductance, and shielding. There are many misconceptions in these areas, and this book is intended to help reduce the confusion about them.

Chapter 2

EMC Fundamentals

2.1 Introduction

The most fundamental fact about EMC is that it is NOT magic, witchcraft, or divine intervention. It is about current flow, electric field and magnetic field coupling, and electromagnetic radiation. The interactions between all the individual system components are often complex and hard to visualize simultaneously. In order to properly understand the various issues, the overall complex problem must be broken into smaller, discrete problems that can be more easily understood. Straightforward scientific and engineering principles can then be applied with success.

This chapter will introduce some basic concepts that are fundamental to EMC problems. A good basic understanding of these fundamentals is necessary before the overall complex problem can be broken into isolated individual problems and solved properly. It is common practice by some EMC engineers to use the 'try-it-and-see' approach. That is, very little consideration is given to root cause analysis (as if the laws of physics might be somehow different for this product). When the product fails EMC requirements, they will try a capacitor here, or a ferrite bead there, or more gasket in another place. Eventually, some combination of things will work together, and the product will pass the EMC requirements. Another approach is the 'shot gun' approach. With this approach, the design engineer includes every possible filter and shielding design options with the hope that something will work. Either of these approaches increase the cost of the product and is a far from optimum design practice.

2.2 Coupling Mechanisms

The two fundamental coupling mechanisms are electric field coupling and magnetic field coupling. The coupling from the noise source to the ultimate radiator could be either electric or magnetic filed coupling or it could be a combination of both. Understanding these coupling mechanisms, and how they might manifest themselves in a product design, is an important part in controlling them.

2.2.1 Electric Field Coupling

Electric field coupling is due to displacement current through a capacitive effect. That is, we do not intend the current to flow in this particular direction, but the natural parasitic capacitance provides a lower impedance path to the current than the intended path provides. Current must always flow in a complete loop, so the loop impedance is the important factor.

For example, Figure 2-1 shows a typical printed circuit (PC) board with a few components on it. A clock buffer drives a trace that passes near a large ASIC/IC with a heatsink above it. As the trace passes the heatsink, there is a parasitic capacitance between the trace and the heatsink. There is also a parasitic capacitance between the heatsink and the clock buffer. (Naturally, there is also parasitic capacitances between the heatsink and the receiver, the shielded enclosure, and all other parts of the system, but for this example, they are small enough to not affect the EMC performance.) The impedance of a capacitor is given by

$$X_c = j\frac{1}{2\pi f C} \qquad (2.1)$$

where
C = capacitance, and
f = frequency.

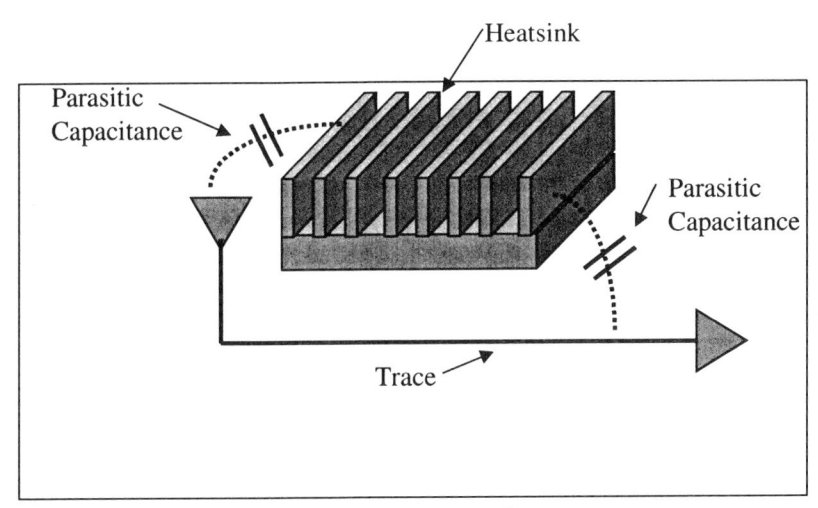

Figure 2-1 PC Board With Components

The amount of capacitance is set by the physical geometry. The impedance, however, will be affected by frequency. Higher frequency harmonics will encounter lower impedance for a given geometry.

Recall that all current must flow in a complete loop back to its source. The intended path in this example is to have the current at all harmonic frequencies leave the clock buffer, travel through the trace to the receiver, and then return to the clock buffer through the ground-reference plane. In this example, however, the parasitic capacitances between the trace and heatsink, and between the heatsink and the clock buffer, provide a lower impedance path than the intended one, causing at least some of the current to flow through the heatsink. While we might not initially care about this return current path, we should recognize that the heatsink is physically larger radiator than the trace. The heatsink is very likely to be a more efficient radiator, especially at high-frequency harmonics, and can cause emissions that were unnecessary and must be contained by improved shielding.

2.2.2 Magnetic Field Coupling

Magnetic field coupling is due to conduction current through an inductive effect. In this case, we do intend the current to flow in a particular direction, but the natural parasitic inductance provides a lower impedance path to the current than the intended path provides. Since current must always flow in a complete loop, the loop impedance is again the important factor.

For example, Figure 2-2a shows a view of two PC board vias that are between two solid planes (such as a power and ground-reference planes). In this example, the same clock buffer from the previous example is used, only this time, the trace is buried between various layers (as shown in Figure 2-2b) to avoid the electric field coupling described in the previous example. The trace usually must change layers at some point (to avoid other traces and devices), and the signal current travels on the via shown in Figure 2-2b.

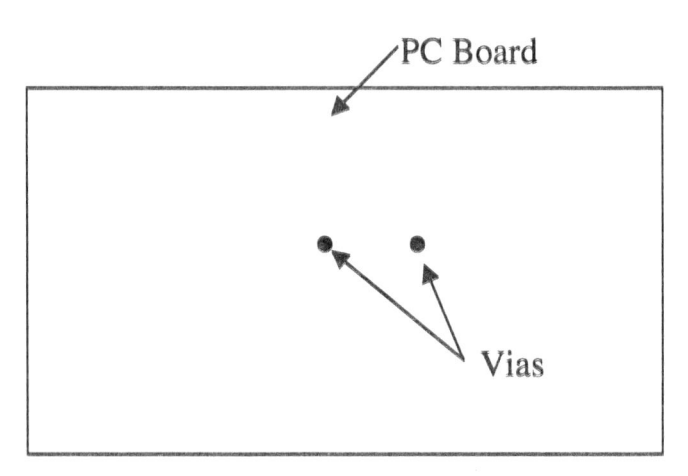

Figure 2-2a PC Board with Vias Example

In this example, the second via is connected to an internal unshielded cable, such as a ribbon cable for a disk drive. The current in the first via causes magnetic flux lines, some of which will intersect the second via, as shown in Figure 2-2c. These magnetic flux lines will induce a current in the second via which will be conducted onto the ribbon cable. This parasitic mutual inductance will have a lower impedance for the higher harmonic frequencies, and will more easily conduct current away from its intentional path and onto potentially harmful unintentional paths (as described in Chapters 6 and 7).

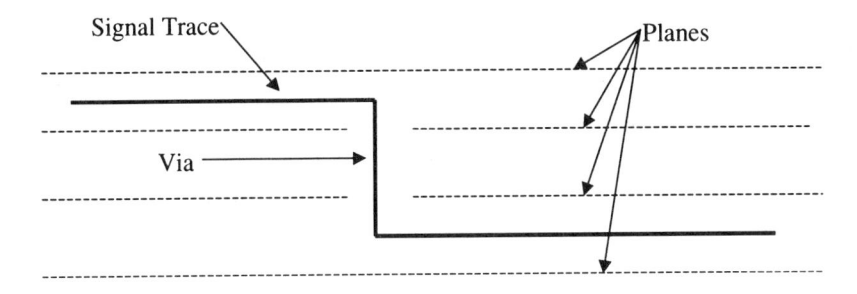

Figure 2-2b PC Board with Vias Example (Inside View)

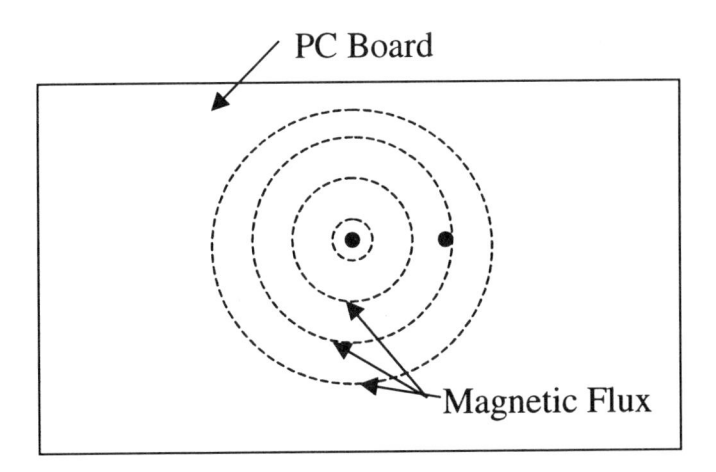

Figure 2-2c PC Board with Vias Example with Magnetic Flux

Recall that all current must flow in a closed loop back to its source. If we suppose the internal ribbon cable has a sufficient parasitic capacitance between the cable and the original clock buffer, then at least some of the current might flow through this path as shown in Figure 2-2d. The combination of the parasitic mutual inductance and parasitic capacitance causes current to flow in the unshielded cable, and radiate onto the structure of the shielded enclosure. This requires the shielded enclosure to have additional shielding, again increasing the cost and complexity of the enclosure.

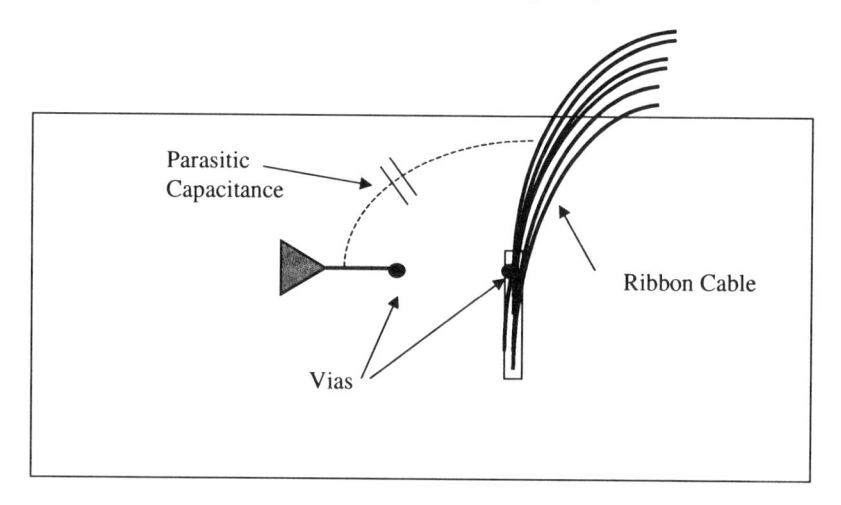

Figure 2-2d Example of Parasitic Return Current Path

2.3 Signal Spectra

The harmonic spectra of the intended signals are a very important EMC design consideration. The fundamental harmonic frequency is seldom the problem frequency. Most EMC problems are from higher frequency harmonics.

2.3.1 Clock Signals' Harmonic Frequencies

From a Fourier analysis, the harmonic frequency content from a simple square wave consists of the fundamental frequency and all the odd harmonics. The amplitude of the individual harmonic frequency is given by

$$A_n = \frac{1}{n} \qquad (2.2)$$

where
n = odd numbered harmonic (1, 3, 5, 7, ...), and
A_n = amplitude of harmonic.

Figure 2-3 shows an example for a 100 MHz square wave. The amplitude is displayed in dB, and it is apparent that the amplitude of the higher frequency harmonics does not decay fast.

This example had no even harmonics because the duty cycle was exactly 50% and the square wave has equal rise and fall times. This is seldom the case in the real world. A slight change in duty cycle will create significant even harmonic amplitude. Even if the duty cycle is exactly 50%, a difference in the rise and fall time will create even numbered harmonics.

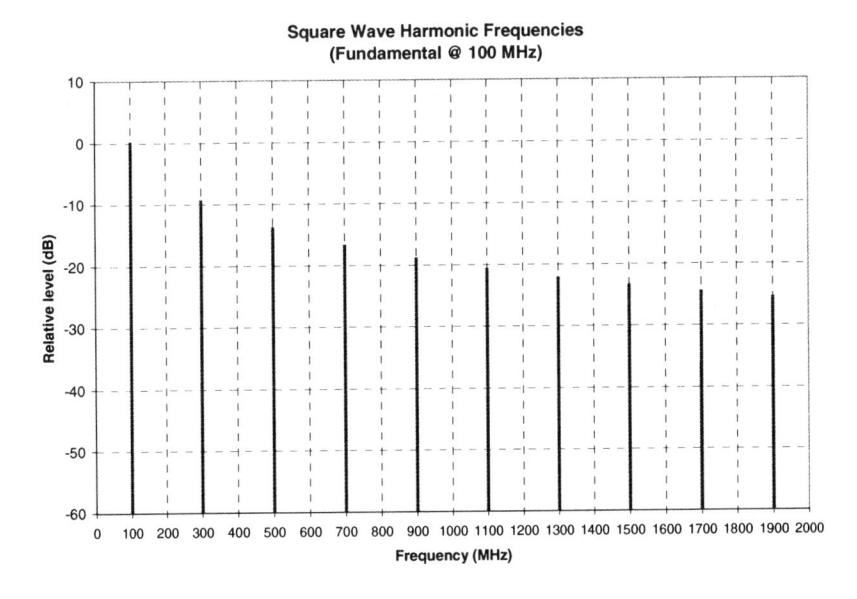

Figure 2-3 Example of Harmonic Content for 100 MHz Squarewave (or 200 Mb/s Squarewave)

16 / PCB Design For Real-World EMI Control

The previous example does not include the effect of rise/fall time for the pulse. In fact, the previous example used both rise and fall times of zero. When the real rise and fall times are included, the higher frequency harmonics are affected. Figure 2-4 shows the envelope of the frequency spectrum of a typical trapezoidal pulse based on the pulse width and the rise and fall times of the pulse. Since higher frequencies tend to radiate more efficiently from traces, etc. and are also able to radiate through smaller openings in the metal enclosure, it is best to keep high-frequency harmonics as low as possible. As Figure 2-4 shows, the amplitude of the pulse spectrum decreases with higher frequency. The spectrum decreases at a rate of 20 dB per decade of frequency above a frequency related to the pulse width, and at 40 dB per decade above a frequency related to the pulse rise/fall time. The slower the rise/fall time, the lower frequency at which the second break occurs, resulting in reduced signal levels at high frequencies. Clearly, the slower the rise and fall times of a pulse, the lower the potential frequency domain harmonic content of that signal.

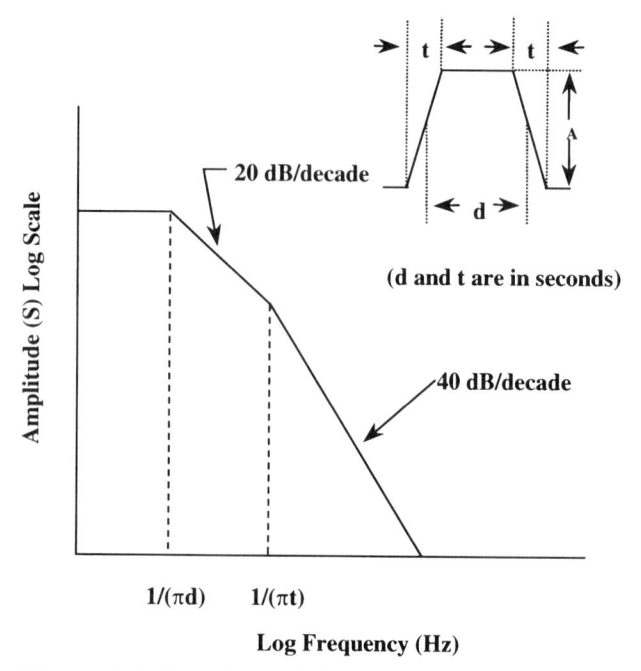

Figure 2-4 Envelope of Spectrum of Trapezoidal Pulse

2.3.2 Hertz vs. Bits-per-Second

There is confusion sometimes about the difference between a (for example) 50 M bit/sec signal and a 50 MHz signal. These are not the same thing, and the fundamental frequency of the 50 M bit/sec signal is not 50 MHz. Figure 2-5 shows an example of a 50 M bit/sec square wave and the fundamental sinewave at 25 MHz. In effect, the squarewave data rate only uses one bit width (which is one half of the full sinewave cycle) to determine the data rate. This means that a 100 M bit/sec pulse will have its odd numbered harmonics at 50 MHz, 150 MHz, 250 MHz, 350 MHz, etc.

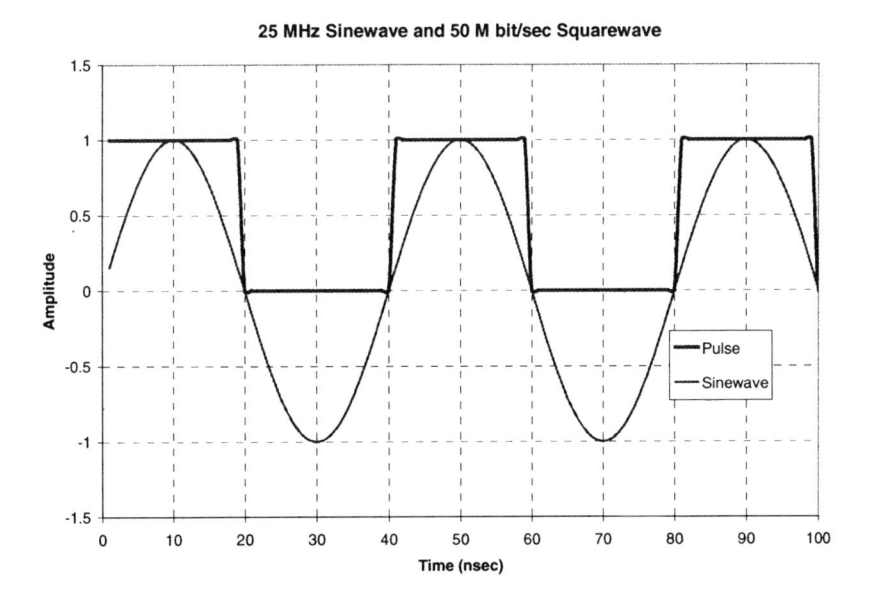

Figure 2-5 Comparison of 25 MHz Sinewave and 50 Mbit/sec Squarewave

18 / PCB Design For Real-World EMI Control

2.3.3 Non-Squarewave Data Signals

Clock signals are typically a square wave, but data and address signals change from moment to moment. The instantaneous spectrum will change with the data as well. The basic harmonic content is similar to the square wave and based on a sinc type function[1]. If a max-hold function is applied to the changing spectrum from a pseudo-random bit stream, an envelope develops as shown in Figure 2-6. The peaks of the harmonic envelope are centered on the squarewave harmonics. At any instant in time, however, the actual harmonic frequency and amplitude will be under the envelope and probably not at the squarewave harmonic.

As mentioned in a previous section, if the duty cycle is not exactly 50%, then the null frequencies in Figure 2-6 will have some non-zero values, depending on the amount of deviation form the 50% duty cycle.

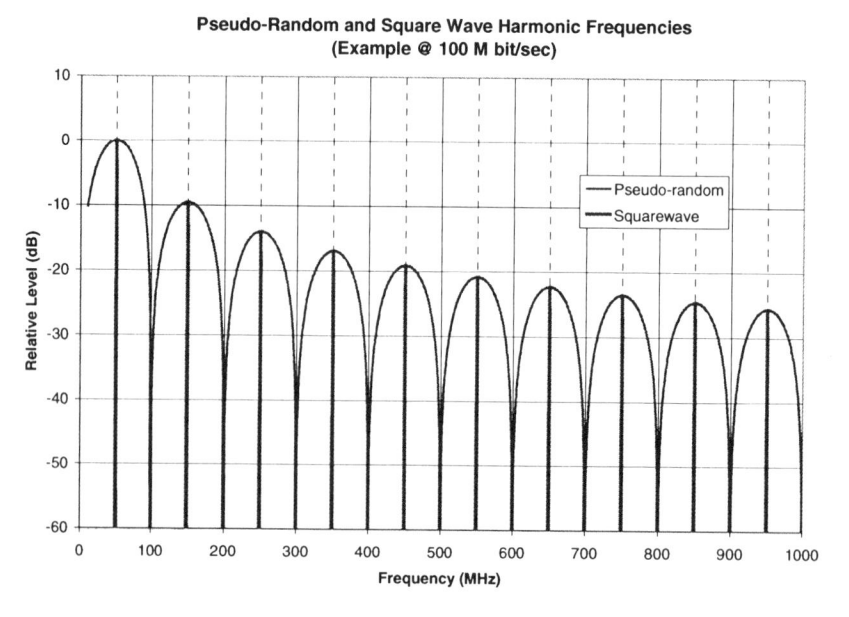

Figure 2-6 Envelope of Spectrum for Pseudo-Random Bit Stream

---

[1] The sinc function is a sin(x)/x function.

2.4 Resonance Effects

Most EMI emissions are not broad band but fairly narrow band. A resonance is excited somewhere in the system. It may be a resonance of an external cable is excited and becomes an efficient radiator, or an internal heatsink is excited and becomes a good radiator, or any number of other things that can become efficient radiators at specific frequencies if excited.

Resonance is either a physical size phenomenon or a circuit based phenomenon. Circuit based resonances are due to capacitive and inductive reactive impedance components being equal and opposite phase. Energy is stored in the capacitor and then in the inductor, repeatedly.

Physical resonance is due to the physical size of the conductor. For example, a wire in free space, when excited at its center, behaves like a dipole antenna. This 'antenna' is an antenna whether or not we intended it to be one! The antenna will be excited most effectively when the size of the wire is one-half the wavelength of the exciting frequency. The physical length of the antenna determines the resonance frequencies. Typically, a straight wire will be resonant at odd numbered half wavelength harmonics.

Resonances can increase the amount of emissions because the final radiator is more efficient. For example, a personal computer product had a front panel LED display. The LEDs needed to be visible to operators and were on the outside of the metal enclosure. By themselves, the LEDs and their associated circuitry were not an efficient radiator at the frequencies (and the harmonics) contained in the circuits. A plastic door was positioned near the LED display area. This plastic door was not part of the metal enclosure and considered not an EMC concern. However, the plastic door latch was built on a metal rod that was about 30 cm long. Even though the metal latch rod was not in physical contact with the LED circuits, the parasitic capacitance and inductance coupled energy onto it, and harmonic signals in the 500 – 600 MHz range from the LED circuits were radiated efficiently! The halfwave resonant frequency for a 30 cm long 'wire' is about 500 MHz, so the metal latch rod was an efficient antenna for harmonics around this frequency. This was an example of where an unexpected resonance increased the emissions significantly.

Cavity resonances are based on the volume of the enclosure/cavity. Boundary conditions force the tangential electric field to be zero on a perfectly conducting metal wall. In an empty rectangular cavity, whenever the cavity size allows an integer number of half wavelengths along any dimension, it will support standing (or stationary) waves, and is resonant. For an empty rectangular cavity, the frequencies where the cavity is resonant is given in [2.1] by

$$f_{mnp} = \frac{1}{2\sqrt{\varepsilon\mu}}\sqrt{\left(\frac{m}{a}\right)^2 + \left(\frac{n}{b}\right)^2 + \left(\frac{p}{c}\right)^2} \qquad (2.3)$$

where:
a, b, and c = length of the sides of the cavity, and
m, n, and p = integer numbers (only one can be zero at a time).

Equation (2.3) only applies to empty rectangular cavities. Typically, electronic or computer products contain a number of circuit boards, internal cables, power supplies, etc. Each of these objects alter the boundary conditions and therefore change the resonant frequencies. In an enclosure with little open space, there is little space to support the standing waves, and the likelihood of internal cavity resonances is low.

2.4.1 Magic and Luck

One of the main reasons that EMC has a reputation for being magical is because of the effects of resonance. These are not generally from straightforward wire or cavity resonances, but because of the interaction between the physical resonances and the circuit based resonances when the parasitic elements are included. As mentioned earlier, the parasitic elements are difficult or impossible to compute with simple closed form equations, and so they are often ignored, but they exist whether they are ignored or not.

For example, in the past, a common design practice was to intentionally not connect the PC board ground-reference plane directly to the metal chassis at all of the PC board mounting screw locations. Often circuit pads were left on the PC board so that one could later install a capacitor, zero-ohm resistor, or ferrite bead as testing required if problems occurred. Unfortunately, this created more problems than it solved since even the zero-ohm resistor adds inductance (impedance) and is a poor connection at high frequencies.

Traditionally, engineers would try different values of capacitor, ferrite, etc. until the emissions were low enough to pass the requirements. Often, while trying various combinations of components, they would observe the signal of interest reduced, but another frequency signal unexpectedly raised above the limits! Unknowingly, they were simply tuning the resonances of the various parasitic elements with their added circuit elements until they were lucky enough to arrive at a combination that allowed the resonances to add or subtract in their favor.

Obviously, this is not the preferred approach. Rather than depend on luck and many hours spent in the EMC laboratory trying various circuit combinations, up front design and consideration of the parasitic circuit elements, the possible resonances, and the overall equivalent circuits will help the designer be successful the first time.

2.5 Potential Emissions Sources

To understand the potential emissions sources in the EMC test chamber, the product under consideration should be enclosed in an imaginary enclosure. Any thing that leaves this imaginary enclosure is a possible emissions source.

2.5.1 Shielded Products

There are two ways for noise energy to escape the shielded enclosure. Energy can escape through openings in the metal enclosure or be conducted through the enclosure on I/O cables and wires.

Openings in the enclosure are typically air vent areas, seams where metal enclosure parts are joined, and any other doors and windows. Noise energy escapes these openings and causes RF current on the outside of the enclosure, cables, etc. in a complex pattern. These currents cause emissions, and depending on the resonant frequencies of the external structure (with cables etc.) will radiate more or less efficiently. The radiation does not necessarily leave the opening, seam, or air vent and radiate directly to the receive antenna. This is why emissions often seem to come from a corner of the enclosure where there is no opening, or even from a metal side panel with no seams. The radiation direction is determined by the entire external system, not the point of leakage.

I/O wires and cable allow internal signals to travel to the outside of the enclosure directly. These signals may be the intended signals, but are often unintended signals that couple onto the I/O traces or connector pins inside the enclosure. Once the unintended signals are on the external cables, they will radiate based on the resonances of the external cable lengths and shapes.

Note that while a straight wire has a resonant frequency (as described earlier), when the wire/cable is bent that resonant frequency changes because the RF current distribution changes. Typically, EMC emissions test procedures require that the external cables be bent in different shapes and their positions changed to optimize the emissions at all frequencies. This requirement is effectively changing the cables resonant frequencies to make them a more efficient radiator.

2.5.2 Unshielded Products

Some products do not include a metal enclosure. This is typical for low cost products where the cost of a metal enclosure is too high. In this case, the PC board can radiate directly, or signals can couple onto the attached cables/wires (as in the shielded case).

Emissions from the PC board can be significant. Energy coupled onto heatsinks and external traces will cause emissions. Often, products that are very cost sensitive will use single or double layer PC board stackup, which eliminates the benefits of solid power and ground-reference planes. The return currents from intentional signals must be routed as traces, and the loop created by the driving trace and the return trace can cause direct emissions.

2.6 Intentional Signal Content

One of the most fundamental facts concerning emissions is that *current* causes emissions, not *voltage*. As engineers, we tend to focus on voltage waveforms and not the current waveforms. This is sufficient if we are dealing in purely resistive circuits, but this is seldom the case. CMOS IC devices have a very different current waveform than the voltage waveform. In addition, if nonlinear devices are included, such as clamping diodes, the amount of current, especially at high frequencies, can be significant.

High frequencies tend to radiate more than low frequencies. This is mostly due to the fact that resonant effects and parasitic elements are more effective at higher frequencies. If the energy is not created to begin with, it cannot be coupled though a parasitic element, nor will it excite a resonance and ultimately cause an emission. The most cost effective way to control emissions is at their source, and to control the spectrum of the intentional signal currents. Chapter 6 will discuss this in much greater detail.

2.7 Summary

EMC is not magic. Effective control of EMC emissions require an understanding of the signal harmonic spectrum of the intentional currents, and how parasitic capacitance and inductance can cause these currents to flow in areas where they were never intended. When intentional currents flow on metal surfaces where they were never intended, natural resonances of those metal parts can significantly increase the emissions.

References

[2.1] Harrington, R.F., "Time-Harmonic Electromagnetic Fields," McGraw-Hill, 1961.

Chapter 3

What is Inductance?

3.1 Introduction

While many of us feel that we understand it, in reality inductance is a commonly misunderstood concept. Inductance is important to EMI/EMC design considerations since it is one of the primary limiting factors in high-frequency design. Whenever there is metal, and current flows though that metal, inductance is present and will affect the current flow. At high frequencies, this intrinsic inductance dominates all components, traces, and metal planes. Capacitors and resistors become inductors.

A complete study of inductance would fill at least one entire book. The purpose of this chapter is to help the reader better understand the concepts of inductance, mutual inductance, and partial inductance as they apply to EMI/EMC design, especially on printed circuit (PC) boards.

3.2 Electromagnetic Induction

When the current in a loop changes with respect to time, the magnetic field associated with that current also changes. As this changing magnetic field cuts through a conductor it induces a voltage in the circuit of that conductor. This occurs whether the magnetic field lines cut through a different conductor or the same conductor as the original current. The voltage induced in a single wire loop is equal to the time rate of change of magnetic flux passing through the wire loop. [3.1] This is described in Faraday's law of electromagnetic induction as:

$$\oint \vec{E} \cdot dl = -\iint \frac{\partial \vec{B}}{\partial t} \cdot d\vec{S} \qquad (3.1)$$

Figure 3-1 shows a simple rectangular loop. If the loop is small compared to the wavelength of the frequency of interest, then it can be assumed that the magnetic flux is constant over the area A, and Equation (3.1) can be reduced to

$$V = -A \frac{\partial B}{\partial t} \qquad (3.2)$$

The amount of voltage induced from a time-varying magnetic field can be found for any geometry using Equation (3.1) and for a simple rectangular loop using (3.2).

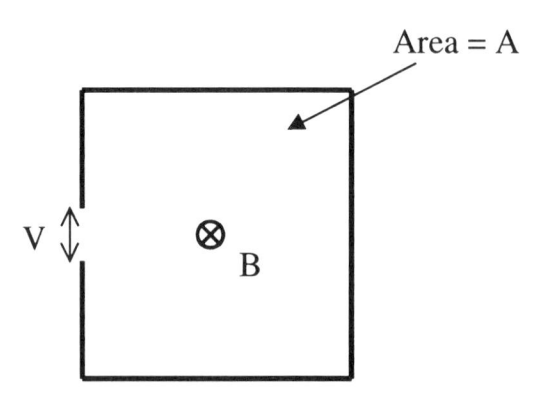

Figure 3-1 Rectangular Loop

3.3 Mutual Inductance

The mutual inductance of real world circuits is often difficult to calculate since the loops are seldom simple geometries, and other metal in the surrounding environment will affect the way the fields behave. If it is assumed that two loops are located in free space (electrically far from other conductors) then the problem is simplified and a reasonable estimate can be made. Under these conditions, the mutual inductance between the two loops is defined as

$$M_{12} = \frac{\int_{S2} \vec{B}_1 \cdot d\vec{S}_2}{I_1} \qquad (3.3)$$

where:
I_1 = the current flowing in loop #1,
B = the magnetic flux created by the current in loop #1, and
S2 = the surface of loop #2.

In Equation (3.3) the magnetic flux from the current in the first loop is integrated across the surface of the second loop to find the mutual inductance.

An alternative form for the mutual inductance may be obtained from vector potential formulations [3.2]. This provides the Neumann form of the mutual inductance of two loops as a double integral around the contours of both loops, in free space, as

$$M = \frac{\mu_0}{4\pi} \oint_{C1} \oint_{C2} \frac{\vec{dl}_1 \cdot \vec{dl}_2}{r} \qquad (3.4)$$

where
r = the distance between contour integration elements in loop #1 and loop #2 as shown in Figure 3-2.

28 / PCB Design For Real-World EMI Control

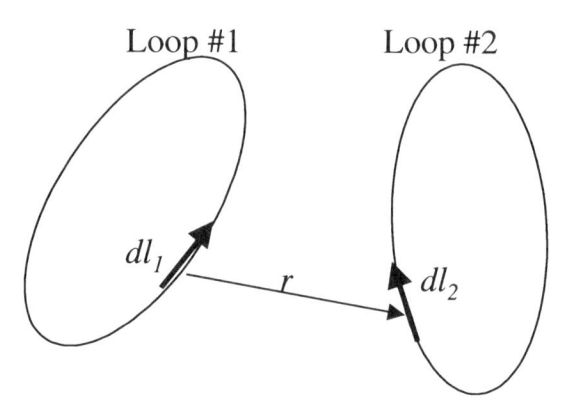

Figure 3-2 Two Arbitrary Loops

The mutual inductance of any general combination of loops can be found from Equation (3.4). For the special case of two circular loops orientated coaxially (as shown in Figure 3-3) and when a << d and b << d, Equation (3.4) can be approximated [3.3] by

$$M = \frac{\mu_0 \pi a^2 b^2}{2(b^2 + d^2)^{3/2}} \qquad (3.5)$$

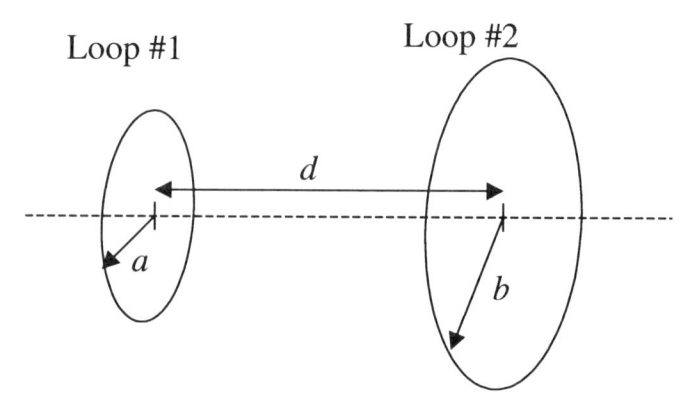

Figure 3-3 Two Loop Coaxial Oriented

Another special case is to find the mutual inductance between two parallel current elements displaced from each other as shown in Figure 3-4. Reference [3.3] provides the Equation (3.6) for the mutual inductance. Distances A through D and overlap distances a through d are indicated in Figure 3-4. The resulting expression is useful to find the mutual inductance between square loop structures, adjacent PC board traces, etc.

$$M = \frac{\mu_0}{4\pi}\left[\ln\frac{(A+a)^a(B+b)^b}{(C+c)^c(D+d)^d} + (C+D) - (A+B)\right] \quad (3.6)$$

Note that while μ_0 is used in all these equations, if the media is anything other than air, then the appropriate permittivity (μ) for that material must be used.

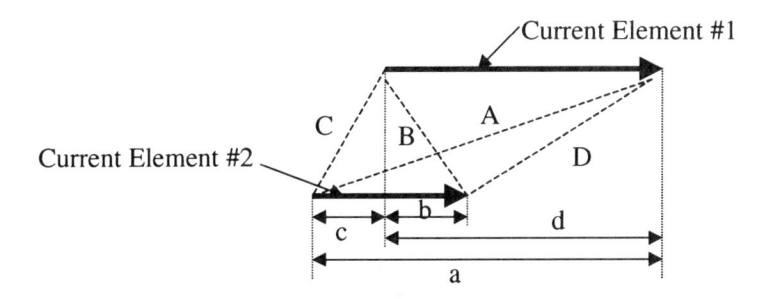

Figure 3-4 Two Parallel and Offset Conductors

3.4 Self-Inductance

Recall from the previous sections that induction occurs when a time-changing current causes magnetic lines of flux to cut through metal conductors. Until now, only the case where these lines of flux caused by a current in one loop cut through the conductors of another loop was considered. These lines of flux will also cut through the conductors of the original loop as well. This gives rise to the loop's self-inductance.

30 / PCB Design For Real-World EMI Control

The self-inductance of an isolated circular current loop in free space can also be found using Equation (3.4) if the two loops are considered to be overlapping. The difference in loop radii is considered to be equal to the radius of the wire of the single loop avoiding a singularity in Equation (3.4). For a simple isolated current loop, where the wire radius r_0 is much smaller than the loop's radius a, then the loops self inductance is approximated as

$$L \approx \mu_0 a \left(\ln \frac{8a}{r_0} - 2 \right) \tag{3.7}$$

If multiple turns of the wire loop are used, then the inductance is simply multiplied by the number of turns to find the total inductance of the number of loops.

For an isolated square loop in free space, the self-inductance can be found from Equation (3.4) using

$$L = \frac{2\mu_0 a}{\pi} \left(\ln \frac{p + \sqrt{1+p^2}}{1+\sqrt{2}} + \frac{1}{p} - 1 + \sqrt{2} - \frac{1}{p}\sqrt{1+p^2} \right) \tag{3.8}$$

where
$p = a/r_0$
a = length of side, and
r_0 = wire radius.

For the case where the wire radius is much smaller than the loop radius ($r_0 \ll a$), Equation (3.8) reduces to

$$L = \frac{2\mu_0 a}{\pi} \left(\ln \frac{2p}{1+\sqrt{2}} - 2 + \sqrt{2} \right) \tag{3.9}$$

Note that for these calculations of self-inductance the contribution by the internal flux within the conductor has been neglected. This term is most important at low frequencies when skin depth is not important and the current is uniformly distributed across the cross section of the conductor. This self-inductance from the internal flux is given as a per-unit-length parameter as

$$L'_{int} = \frac{\mu_0}{8\pi} \qquad (3.10)$$

This term is then multiplied by the loop length to find the total contribution of the internal flux.

For a single turn rectangular loop in free space (Figure 3-5), the self-inductance can be found from

$$A = h \ln\left(\frac{h + \sqrt{h^2 + w^2}}{w}\right)$$

$$B = w \ln\left(\frac{w + \sqrt{h^2 + w^2}}{h}\right) \qquad (3.11)$$

$$C = h \ln\left(\frac{2h}{a}\right) + w \ln\left(\frac{2w}{a}\right)$$

$$L = \frac{\mu_0}{\pi}\left(-2(w+h) + 2\sqrt{h^2 + w^2} - A - B + C\right)$$

where
w = the width of the rectangle (wide dimension)
h = the height of the rectangle (short dimension), and
a = the wire radius.

For a single equilateral triangular loop in free space (Figure 3-6), the self-inductance can be approximated from

$$L \approx \frac{3\mu_0 s}{2\pi}\left(\ln\frac{s}{a} - 1.405\right) \qquad (3.12)$$

where

32 / PCB Design For Real-World EMI Control

s = the length of one side of the triangle, and
a = the wire radius,

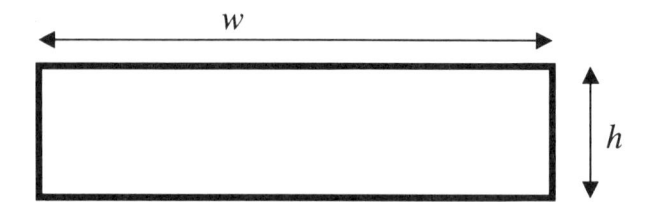

Figure 3-5 Single Turn Rectangular Loop

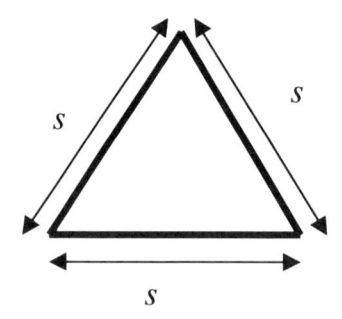

Figure 3-6 Single Equilateral Triangular Loop

For a single isosceles triangular loop in free space (Figure 3-7), the self-inductance can be approximated from

$$A = 2(b+c)\sinh^{-1}\left(\frac{b^2}{\sqrt{4b^2c^2 - b^4}}\right)$$

$$B = 2c\sinh^{-1}\left(\frac{2c^2 - b^2}{\sqrt{4b^2c^2 - b^4}}\right) - (2c+b) \qquad (3.13)$$

$$L \approx \frac{\mu_0}{2\pi}\left(2c\ln\left(\frac{2c}{a}\right) + b\ln\left(\frac{2c}{a}\right) - A - B\right)$$

where
c = the length of the equal sides of the triangle,
b = the length of the base of the triangle, and
a = the wire radius.

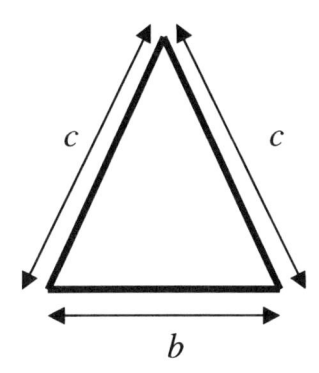

Figure 3-7 Single Isosceles Triangular Loop

3.4.1 Self-Inductance per Unit Length

There are a number of special structures that are commonly of interest to EMI/EMC engineers. Many of these special structures lend themselves nicely to per-unit-length parameters, such as pair of wires, traces over ground-reference planes, etc. This section will provide estimates of the self inductance for a few special structures.

For a pair of wires in free space (Figure 3-8) with the separation much greater than the wire radius ($a \ll d$), the self-inductance can be approximated from

$$L \approx \frac{\mu_0}{\pi} \cosh^{-1}\left(\frac{d}{2a}\right) \qquad (3.14)$$

where
a = the wire radius, and
d = the separation from the center of the wires.

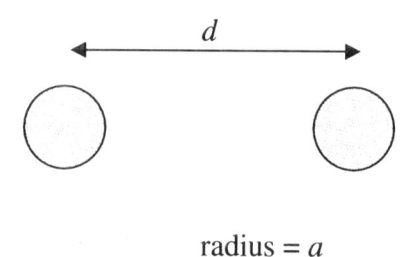

Figure 3-8 Pair of Wires in Free Space

For a wire over a metal plane in free space (Figure 3-9) with the separation much greater than the wire radius ($a \ll h$), the self-inductance can be approximated from

$$L \approx \frac{\mu_0}{2\pi} \cosh^{-1}\left(\frac{h}{a}\right) \qquad (3.15)$$

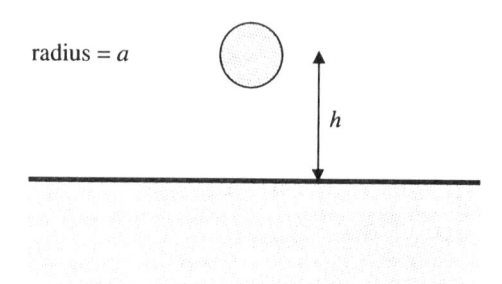

Figure 3-9 Wire Over a Metal Plane

For a flat trace over a metal plane in free space (Figure 3-10), with the width of the trace much greater than the height above the metal plane ($h \ll w$) and the height above the metal plane greater than the thickness of the trace (h > t), the self-inductance can be approximated from

What is Inductance? / 35

$$L \approx \frac{\mu_0 h}{w} \quad (3.16)$$

where
w = the width of the trace,
h = the height of the trace above the metal plane, and
t = the thickness of the trace.

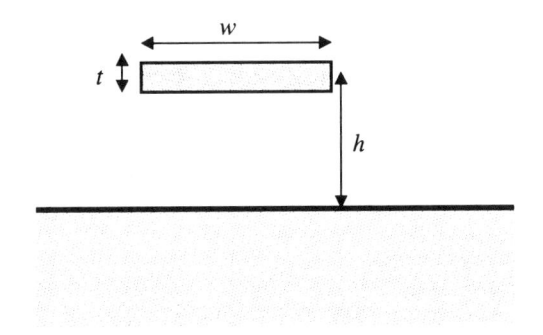

Figure 3-10 Trace Over a Metal Plane

For two flat traces in free space (Figure 3-11), with the width of the trace much larger than the separation between traces ($h \ll w$) and the separation of the traces is greater than the thickness of the trace ($h > t$), the self-inductance can be approximated from

$$L \approx \frac{\mu_0 h}{w} \quad (3.17)$$

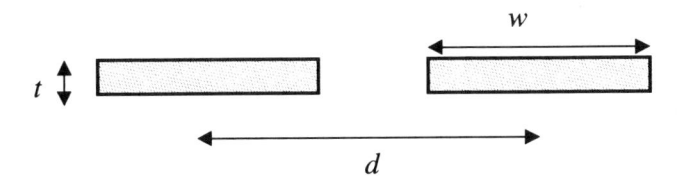

Figure 3-11 Two Flat Traces in Free Space

For two flat co-planar traces in free space (Figure 3-12), with the width of the trace much smaller than the distance between the center of the traces (w << d) and the width of the traces is greater than the thickness of the trace (w > t), the self-inductance can be approximated from

$$L \approx \frac{\mu_0}{\pi} \cosh^{-1}\left(\frac{d}{w}\right) \qquad (3.18)$$

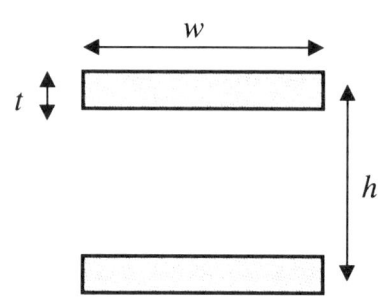

Figure 3-12 Two Flat Co-planar Traces

3.5 Partial Inductance

The definition of inductance requires a current flowing in a loop. *Without a complete loop, there cannot be inductance.* Practical considerations, however, lead us to discuss the inductance of a part of the overall current loop, such as the inductance of a capacitor. This idea of discussing the inductance of only a portion of the overall loop is called partial inductance [3.4]. Partial inductances can be combined to find the overall inductance using Equation (3.19).

$$L_{total} = L_{p1} + L_{p2} + L_{p3} + L_{p4} - M_{p13} - M_{p24} \qquad (3.19)$$

The concept of partial inductance is especially useful when the physical geometry is complex, or when the current is not uniform throughout the cross section of the metal. For example, if a metal bar (as shown in Figure 3-13) is considered to be small enough so that the current is constant throughout the cross section, then the bar could be replaced with an equivalent circuit of a simple series resistor and inductor. The resistance would be given as

$$R = \frac{l}{\sigma w t} \text{ ohms} \tag{3.20}$$

where
l = the length of the bar,
w = the width of the bar
t = the thickness of the bar, and
σ = the conductivity of the material.

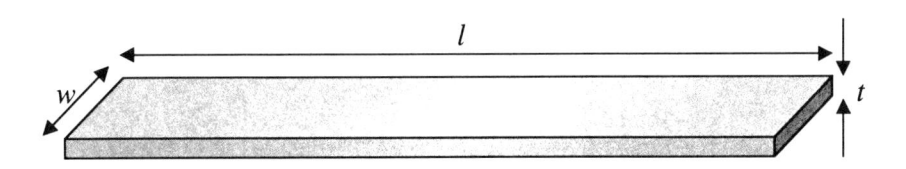

Figure 3-13 Metal Bar for Partial Inductance Calculations

A similar calculation [3.4-3.6] will provide the partial self-inductance, L_{pii}

$$\frac{L_{pii}}{l} = \frac{2\mu}{\pi}\left\{\frac{\omega^2}{24u}\left[\ln\left(\frac{1+A_2}{\omega}\right)-A_5\right]\right.$$

$$+\frac{1}{24u\omega}\left[\ln(\omega+A_2)-A_6\right]+\frac{\omega^2}{60u}(A_4-A_3)$$

$$+\frac{\omega^2}{24}\left[\ln\left(\frac{u+A_3}{\omega}\right)-A_7\right]+\frac{\omega^2}{60u}(\omega-A_2)$$

$$+\frac{1}{20u}(A_2-A_4)+\frac{u}{4}A_5-\frac{u^2}{60}\tan^{-1}\left(\frac{\omega}{uA_4}\right)$$

$$+\frac{u}{4\omega}A_6-\frac{\omega}{6}\tan^{-1}\left(\frac{u}{\omega A_4}\right)+\frac{A_7}{4}-\frac{1}{6\omega}\tan^{-1}\left(\frac{u\omega}{A_4}\right)$$

$$+\frac{1}{24\omega^2}\left[\ln(u+A_1)-A_7\right]+\frac{u}{20\omega^2}(A_1-A_4)$$

$$+\frac{1}{60u\omega^2}(1-A_2)+\frac{1}{60u\omega^2}(A_4-A_1)$$

$$+\frac{u}{20}(A_3-A_4)$$

$$+\frac{u^3}{24\omega^2}\left[\ln\left(\frac{1+A_1}{u}\right)-A_5\right]$$

$$+\frac{u^3}{24\omega}\left[\ln\left(\frac{\omega+A_3}{u}\right)-A_6\right] \quad (3.21)$$

$$+\frac{u^3}{60\omega^2}\left[(A_4-A_1)+(u-A_3)\right]$$

Where

$$A_1 \equiv (1+u^2)^{1/2}$$
$$A_2 \equiv (1+\omega^2)^{1/2}$$
$$A_3 \equiv (u^2+\omega^2)^{1/2}$$
$$A_4 \equiv (1+u^2+\omega^2)^{1/2}$$
$$A_5 \equiv \ln\left(\frac{1+A_4}{A_3}\right)$$
$$A_6 \equiv \ln\left(\frac{\omega+A_4}{A_1}\right)$$
$$A_7 \equiv \ln\left(\frac{u+A_4}{A_2}\right)$$
$$u \equiv l/w, \text{and}$$
$$\omega \equiv t/w.$$

While (3.21) appears to be quite complex, it is straightforward to implement in a computer program. This calculation can be extended to a larger bar [3.7] that has non-uniform current across the cross section by dividing the large bar into smaller bars, each with a uniform current distribution (Figure 3-14). This approximates the current distribution across the large bar as a step distribution, and as long as there are enough smaller bars, the step size between adjacent small bars is low enough for good accuracy. The voltage across each small bar is

$$V = R_i I_i + j\omega L_{pi} I_i + j\omega \sum_{\substack{k=1 \\ k \neq i}}^{N} (L_{pik} I_k) \qquad (3.22)$$

where L_{pik} is the mutual partial inductance between the I-th and k-th small bars and which contains the relative location of the small bars.

Formulas for these mutual partial inductances between rectangular conductors with uniform current distribution are given in [3.4-3.6] with an excellent introduction to partial inductances in [3.4]. The sum of all the small bar currents equal the total current in the large bar. The voltage across each small bar is equal to each other and also equal to the voltage across the large bar. Figure 3-15 shows a schematic representation of this combination of partial inductances.

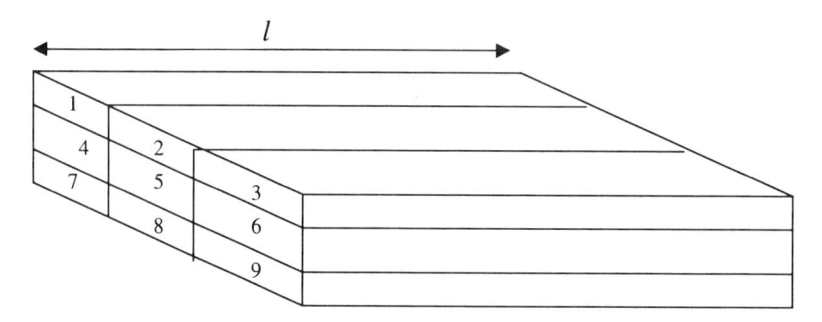

Figure 3-14 Combining Individual Bars to Create Larger Structures

3.6 Summary

The basic principle that inductance requires current to flow in a loop is an important concept to understand. This is not unreasonable since current *must* flow in a loop. The size of the current loop determines the amount of inductance.

Inductance is a basic building block in electronic circuits. That is, as soon as metal conductors are used, and current flows through them, inductance exists. This inductance becomes the limiting factor in all high-frequency circuits. When capacitors are used as filter elements, the natural inductance associated with the current flowing though the capacitor limits the frequency range where the capacitor is an effective filter component.

Partial inductance is a useful concept, since with partial inductances one can discuss the contribution to the total inductance of a single part of the loop. An example is the PC board via connecting between different layers on the PC board, the metal stand-off post between the PC board and the chassis, and traces on the PC board connecting between filter components. Each of these metal structures can be analyzed to find their partial inductances, and the results can then be combined to find the total inductance.

This has been a very brief introduction to inductance. A much more complete study of this subject is available in the references.

References

[3.1] J.D. Kraus and K.R. Carver, *Electromagnetics*, 2nd Edition, McGraw-Hill, 1973

[3.2] F.M. Tesche, M.V. Ianoz, and T. Karlsson, *EMC Analysis Methods and Computational Models*, Wiley-Interscience, 1997

[3.3] Ramo, S., J.R. Whinnery, and T. Van Duzer, *Fields and Waves in Communication Electronics*, 2nd Edition, Wiley, New York, 1965

[3.4] A.E. Ruehli, "Inductance Calculations in a Complex Integrated Circuit Environment," *IBM J. Research and Development*, 16, pp 470-481, 1972

[3.5] F.W. Gover, *Inductance Calculations*, Dover Publications, NY, 1946

[3.6] C. Hoer and C. Love, "Exact Inductance Equations for Rectangular Conductors with Applications to More Complex Geometries," J. Res Nat. Bureau of Standards-C. Eng Instrum., 69C, pp 127-137, 1965

[3.7] C.R. Paul, *Analysis of Multiconductor Transmission Lines*, Wiley, 1994

Chapter 4

The Ground Myth

The term "ground" is probably the most misused and misunderstood term in EMC engineering, and in fact, in all of circuit design. Ground is considered to be a zero potential region with zero resistance and zero impedance at all frequencies. This is just not the case in practical high-speed designs. The one thing that should be remembered whenever the term "ground" is used, is that *"Ground is a place where potatoes and carrots thrive"!* By keeping this firmly in mind, many of the causes of EMC emissions problems would be eliminated.

The term "ground" is a fine concept at DC voltages, but it just does not exist at the frequencies running on today's typical boards. All metal has some amount of resistance, and even if that resistance was near zero ohms, the current flowing through a conductor in a loop creates inductance. Current through that inductance results in a voltage drop. This means that the metal ground plane/wire/bar/etc. has a voltage drop across it, which is in direct contradiction with the intention and definition of ground.

4.1 Where Did The Term "Ground" Originate?

The term "ground" was first used in electronics in the early days of the telegraph. Recall the old cowboy movies where the bad guy shoots down the single wire at the top of the telegraph pole so that the bank he just robbed cannot tell the next town and have them send a posse after him. Note there is only one wire, but in our very first circuits class it is emphasized that current must flow in a loop. It flows down the wire from the sender to the receiver, and then it must return. Back in the days of the telegraph, wire was expensive. It was quickly discovered that if one side of the sending circuit and one side of the receiving circuit were connected to the earth (ground), then only half the wire was needed and the system worked fine. The send and receive ends were "grounded" as in Figure 4-1 and the telegraph circuit was complete.

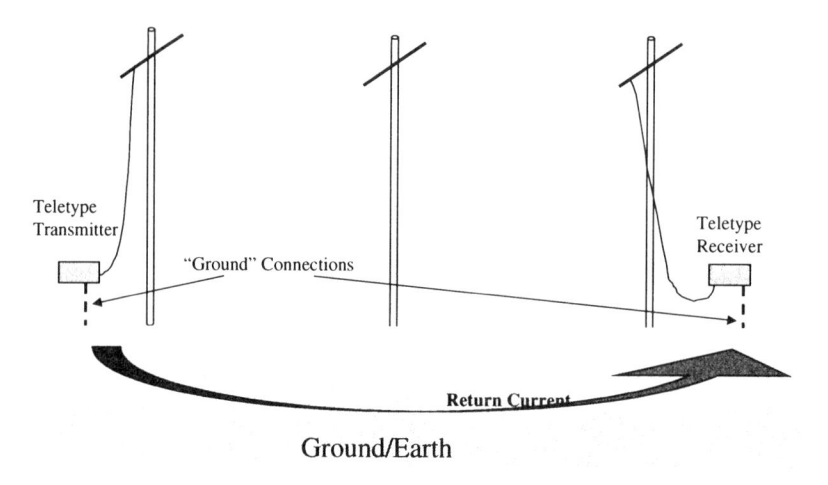

Figure 4-1 --- Ground Current for Very Low Speed Communications

This was an excellent solution when the pulse signal rate was a maximum of approximately 5 pulses per second. Regardless of the conductivity of the earth, the signal got through. Obviously, this is completely unacceptable as data rates increased because the impedance of the earth increases quickly as frequency is increased. Now the entire current path must be considered. This includes not only the direct transmit path (or signal path), but also the return current path (or the signal return current).

The confusion about "ground" is increased with our existing schematic standard symbols. The symbols in Figure 4-2 indicate normal circuit "ground", and even chassis "ground", all of which are assumed to be connected together. While the signal and power return current paths are not shown in typical circuit schematics for simplicity and clarity, these return current paths nonetheless exist, as shown in Figure 4-3. With the return current paths indicated, the current obviously flows in a complete loop.

Even the more complete schematic representation in Figure 4-3 is misleading, since the current flows in three dimensional space and not only in the two dimensional circuit representation in the schematic. Figure 4-4 shows a simple Integrated Circuit (IC) driver and receiver on a printed circuit board (PCB) when the signal line is a microstrip trace[1] using the metal plane as the return current path. Note that the return current flows in the metal plane, and any break in that return current path will interrupt the current loop. Unfortunately, a break in this return current path through the metal plane is common, and this results in the return current taking an unpredictable and most likely problematic path that can lead to interference with other circuits, or EMI.

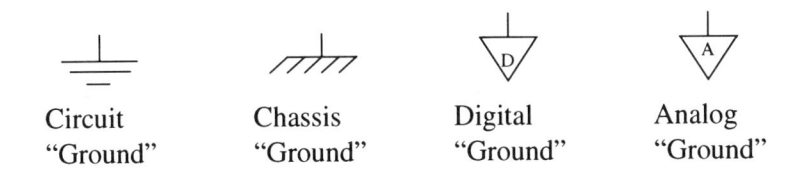

| Circuit "Ground" | Chassis "Ground" | Digital "Ground" | Analog "Ground" |

Figure 4-2 -- Typical "Ground" Symbols

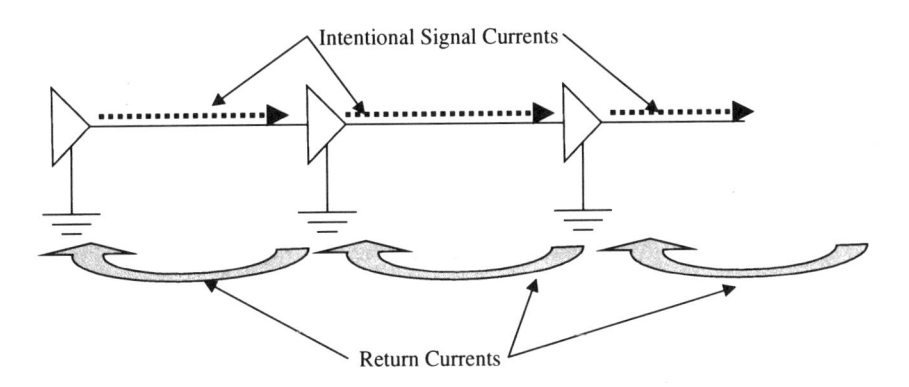

Figure 4-3 -- Schematic With Return Paths Shown

---

[1] A microstrip is a single trace placed over a reference (ground) plane.

46 / PCB Design For Real-World EMI Control

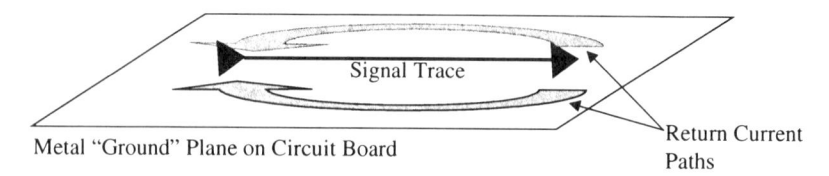

Figure 4-4 -- IC Driver on PC Board with "Ground" Plane (Reference)

4.2 What Do We Mean When We Say "Ground"?

The term "ground" is often used by different engineers to mean different things. Sometimes "ground" is used to mean signal return/reference. Sometimes "ground" is used to mean power return/reference. Sometimes "ground" is used to mean chassis reference, and still other times "ground" means the safety connection to the building metal and ultimately connected to the earth. All of these double meanings are valid connections, but by labeling them all with the same name ("ground") and a multitude of symbols the specific function in the design is totally confused and often misused. A more descriptive name should be used so to eliminate any confusion during design discussions, and to help the designer think more clearly about the function intended.

4.2.1 Signal Reference

One of the most common uses of the term "ground" is to describe the signal reference, or signal return current path. In a simple low speed circuit board, the return current path may be an additional trace, so that the signal current flows 'out' to the receiver on the signal traces and flows 'back' to the transmitter/driver on the signal return traces. Figure 4-5 shows an example of this PCB strategy. Using this design strategy, the return current path is explicitly designed, and care can be taken to ensure this path is not interrupted or routed in a fashion that interferes with other circuits, or picks up noise from other circuits.

Most modern PCBs require signal rates in the 10's, 100's or 1000's of Megabits per second. Functionality concerns require a more controlled signal transmission line. The physical PCB

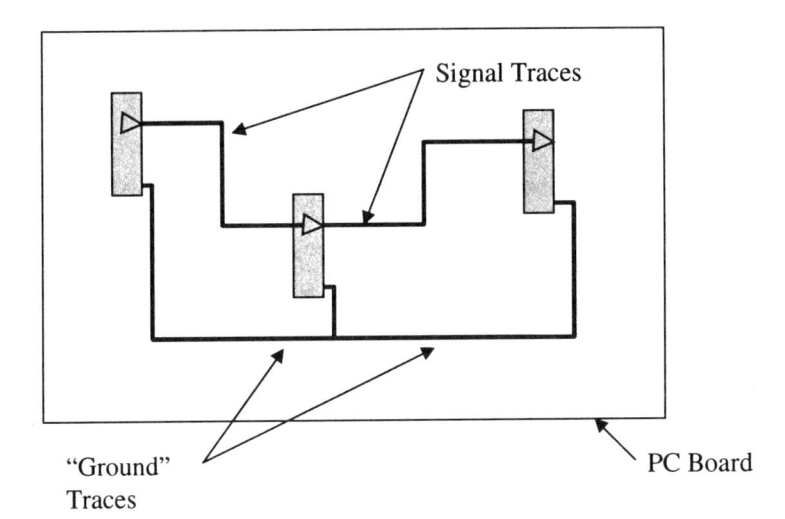

Figure 4-5 – PCB With Separate Traces for Signal and Signal Return

arrangement for a microstrip line and a strip line configuration are shown in Figure 4-6. The signal current flows in the trace and the return current flows in the plane(s). Because of the close coupling between the signal trace and the reference plane(s), the return current flows directly beneath (or above) the signal trace in the reference plane(s).[2] For the simple microstrip example in Figure 4-6, the return current will be (mostly) under the microstrip on the top side of the reference plane. For the symmetric stripline shown in Figure 4-6, the signal return current uses both planes equally. As long as no discontinuities in the reference plane exist, such as a break in the plane, a via transition, etc., the return current remains closely coupled to the signal current and an effective transmission line is created.

Another common PCB configuration is the asymmetric stripline trace shown in Figure 4-7. This configuration is often used when multiple layers are required in the PCB stackup. The majority of the signal return current from an asymmetric stripline returns on the

---

[2] Note: most of the return current flows directly beneath/above the signal trace, but some of the return current spreads out to find the path of least inductance. Please see Section 6.6 for more discussion on this subject.

48 / PCB Design For Real-World EMI Control

closest reference plane. Again, because of the close coupling between the signal trace and the reference plane(s), the return current flows directly beneath (or above) the signal trace in the reference plane. As long as no discontinuities in the reference plane exist, such as a break in the plane, a via transition, etc., the return current remains closely coupled to the signal current and an effective transmission line is created.

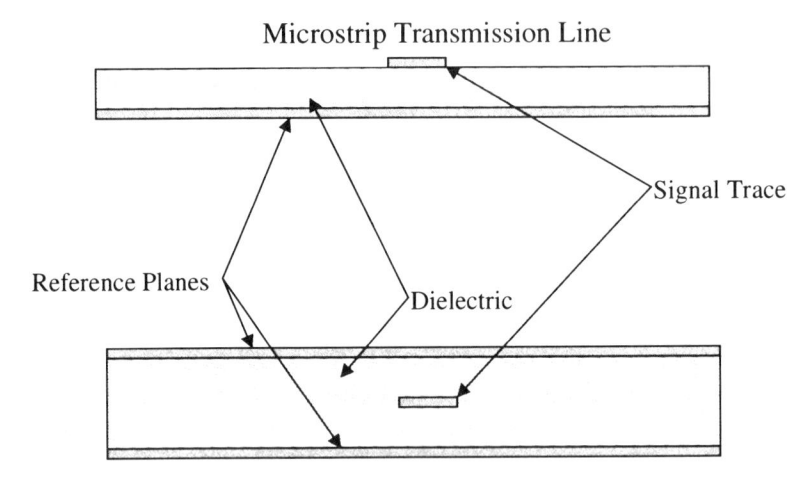

Figure 4-6 – Symmetrical Stripline and Microstrip Line Configuration

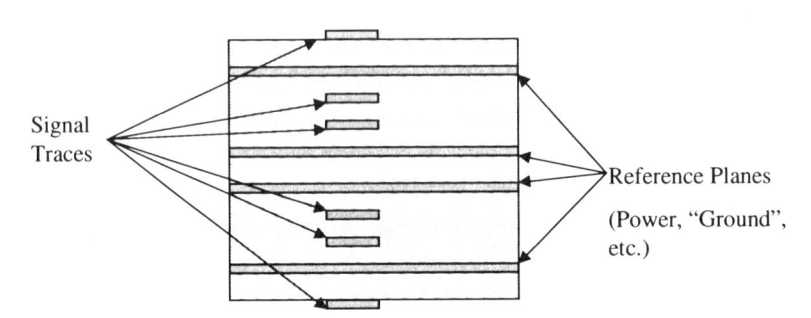

Figure 4-7 – Asymmetrical Stripline and Microstrip Line Configuration

4.2.2 Power Reference

Another common meaning of the term "ground" is the power current return. Typically, no distinction is made for an IC between signal current return and power current return. ICs require power to drive signal current down the signal trace, as well as to operate circuits, logic gates, etc. inside the IC itself. This internal power current must return to the power supply through the power reference connection. The current path for this purpose does not follow the signal traces, but flows in a completely different path.

In the case of a simple low speed PCB, the power connection is clear, and the power return current path straightforward, as shown in Figure 4-8. Power is supplied to the driver IC, some of the current flows outward on the signal trace, and the remainder flows directly back to the power supply and decoupling capacitor.

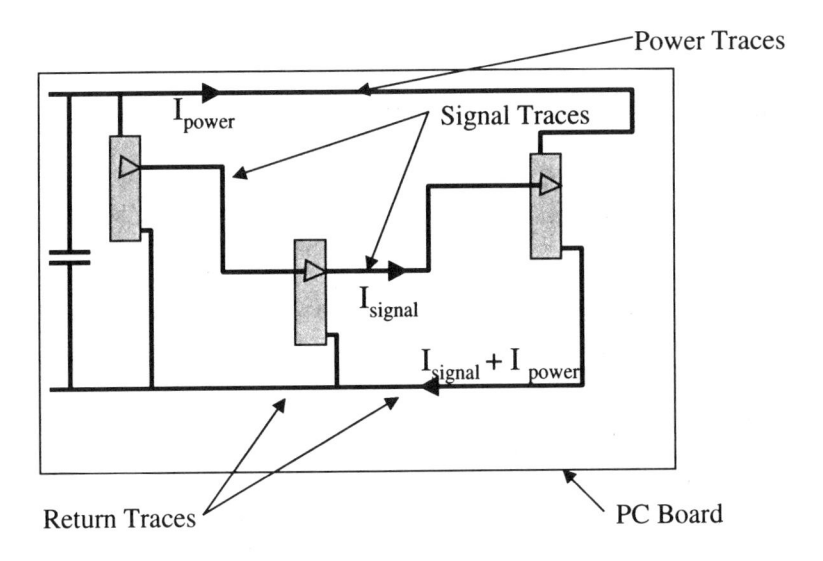

Figure 4-8 -- PCB With Separate Traces for Signal, Power, and Return

When the signaling speeds require more controlled transmission lines, such as microstrip lines and striplines, then the signal reference planes are often used as a power supply plane and a "ground" or power reference plane[3]. This means that the signal reference plane could either be a power plane or a power reference plane as well as the signal reference plane. By using the term "ground" for signal return and power return, when the signal line is referenced to the power plane, things become very confusing. How can a single plane be both power and ground?

This confusion can sometimes be reduced by referring to the power plane as "AC ground", because the decoupling capacitors connect the power plane and the "ground" plane together at high frequencies. While this is true over a limited range of frequencies, the inductance of the capacitor, and more importantly, the inductance of the vias, pads, and traces connected to the capacitor limit the high-frequency performance of the capacitor. This results in a non-zero impedance between the power and "ground" plane, and a non-"AC ground".

Even if all signals and power are referenced against the same plane, the signal return currents and the power return currents flow in different paths. This is illustrated in Figure 4-9. Since the power for the IC is likely to be supplied from both the power supply and the local decoupling capacitors, some of the return current must flow to each location (to complete the required current loop(s)).

4.2.3 Chassis Reference

Most products use a shielded metal enclosure to house the circuit cards, etc. All internal power references and all internal signal references should be connected to this metal chassis close to the connectors where all the external cables leave the board. The metal chassis is very important to the overall emissions performance of the system. The most common cause of EMI emissions is unintentional common-mode currents on external cables and cable shields. These currents result from a voltage difference between the cable (or cable shield) and the chassis metal. From the external emissions point of view, the chassis is the ultimate 'ground', or reference for these unwanted voltages. This means that the primary strategy to control external emissions is to reduce the unwanted voltages between the cable (or cable shield) and the chassis. The consideration is similar

---

[3] Please refer to Chapter 5 for more discussion on return current for transmission lines.

for the unshielded and shielded cables, but there are slight differences for each of these cases.

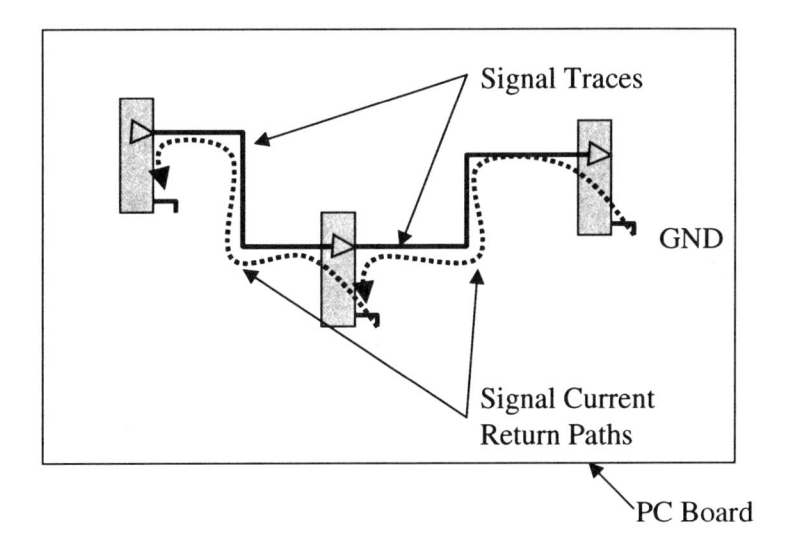

Figure 4-9 -- Ground Plane Return Currents for Power and Signal

4.2.4 Unshielded Cables

An unshielded cable has the conductors directly connected to the internal circuits, typically through a connector. Since these conductors carry intentional signals, the frequency content and/or the signal level must be low enough that these intentional signals do not cause emissions themselves.

There are a number of ways that an unintentional signal, or noise might be also present on the same conductors, but at a voltage level less than the intentional signals. The intentional I/O driver might have internal noise coupled onto the I/O signal line. Fields within the metal enclosure might couple onto the trace leading to the I/O connector, or even directly to the connector pins. There might be crosstalk coupling between high-speed clock or bus lines onto the I/O trace. There are other possibilities as well. Regardless of how these unintentional signals may couple onto this I/O line, a filter is required to reduce the unintentional noise voltage. Typically, this

52 / PCB Design For Real-World EMI Control

noise voltage must be below 100 microvolts (0.0001 volts) to insure passing commercial regulatory limits (using a simple resonant dipole calculation) [4.1].

Chapter 9 will discuss in detail how to design effective filters. One of the most common strategies for filter design includes a capacitor between the signal line and the circuit board reference plane, as shown in Figure 4-10. Naturally, this filter must be designed so that the intentional signal is allowed through, and the unintentional signals are attenuated. The important consideration is for the unintentional signal on the I/O line to be attenuated relative to the chassis, not just relative to the circuit board reference. The filters normally used are mounted on the circuit board for convenience and lower cost, and all attenuate relative to the circuit board reference and not directly to the chassis reference. The impedance of the connection between the circuit board reference and the chassis reference allows a voltage drop across this impedance and therefore reduces the effectiveness of this I/O line filter[4]. Figure 4-11 shows a schematic drawing of this connection impedance.

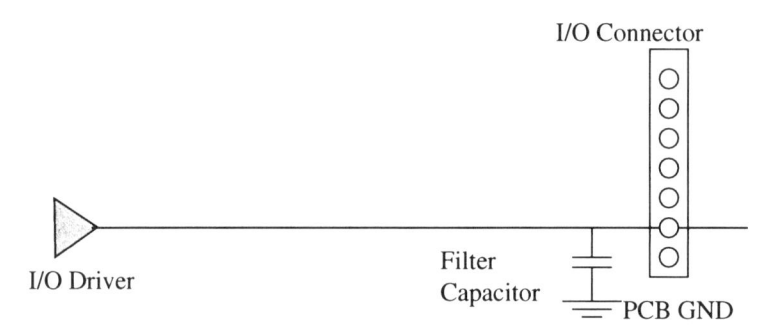

Figure 4-10 --- I/O Filter Schematic

Since the impedance of the connection between the circuit board reference and the chassis reference is part of the filter design (unintentionally), this impedance must be minimized to ensure the filter will do the intended job effectively. At high frequencies, the DC conductivity is not an issue. The concern is due to the inductance of the PCB reference and chassis connection. Even a

---

[4] The connection between the PC board and the chassis is discussed later in this chapter.

perfect conductor has inductance, and therefore has impedance. Once a noise current is flowing through an inductance there will be a voltage drop. This noise voltage can then effectively drive the I/O lines relative to the chassis.

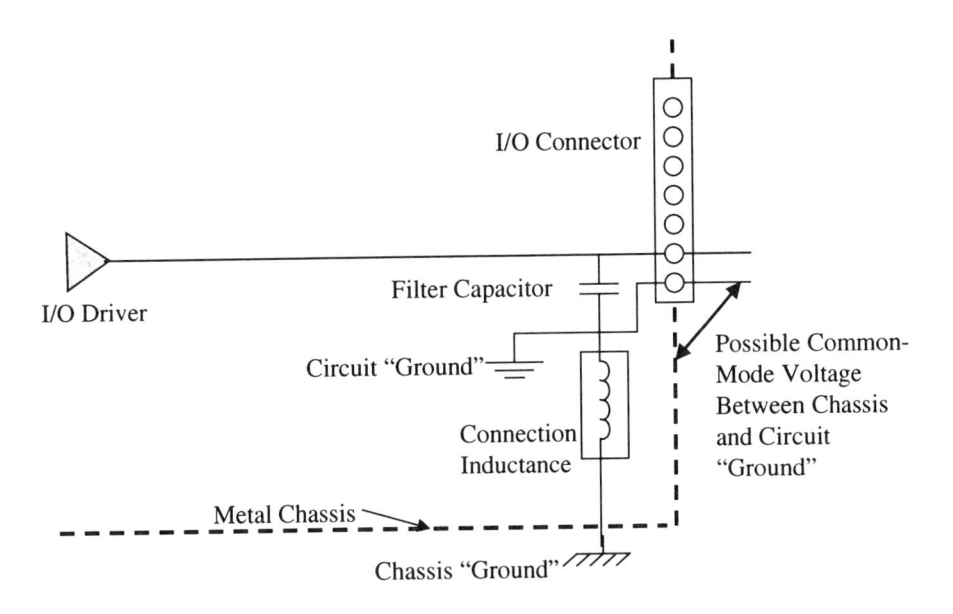

Figure 4-11 --- I/O Filter Schematic with Connection Inductance

The loop inductance for various distances between the I/O connector and the metal post is dependent upon the loop area, not the lateral distance. The inductance, and thus the impedance, increases very rapidly if the standoff metal posts are not positioned close to the connector area.

Another concern is the size of the connection between the board reference and the chassis. Part of the loop is the reference plane on the board, and another part is the chassis itself. These conductors are wide and have minimal partial inductance. The contact between the board reference and the chassis is typically much smaller and its partial inductance contributes significantly to the total inductance, and dominates to the impedance of the total path. This contact interface is where most of the noise voltage that drives the cable relative to the chassis is generated.

At high frequencies, the currents can only flow on the surface of the conductors. This is known as "skin-effect". The skin-effect limits the area the current can flow in, which increases the effective partial inductance. From this analysis, it is apparent that the size of the connection post is also important. The posts should be as short and as thick as possible. This will minimize the partial inductance of the post and therefore the total loop inductance (which consists of the various partial inductances). One common design strategy uses connectors with metal tabs that contact directly to the chassis and also to the board reference plane. When sufficient numbers of tabs are used, the inductance is minimized. If too few tab connections are used, the connection impedance is not lowered, and the noise voltage relative to the cable is again created.

4.2.5 Shielded Cables

When the intentional I/O signal is at a high-frequency or data rate, a shielded cable is required to ensure the intentional signal will not cause emissions over the required limits. This is an effective strategy to keep the intentional signal from causing radiated emissions as long as the cable shield is connected to the chassis with a low inductance/impedance path.

Some shielded cables use a pigtail strategy to connect the shield of the cable. With this strategy, the cable shield is terminated some distance from the connector and a thin wire is connected to some metal part of the connector, and ultimately to the chassis. In some applications, this wire is connected to one of the connector pins, then to the circuit board reference. As described above, the thin wire conductor greatly increases the partial inductance of the connection as compared to a short, fat wire, and therefore the impedance is high. When the thin wire pigtail is connected directly to the metal of the chassis, there is still significant impedance between the cable shield and the chassis shield. Any currents flowing on the inside of the cable shield must also flow across this pigtail impedance. This effectively defeats any shield advantage, and causes a voltage between the chassis and the cable shield, the very thing the shield was being used to avoid.

4.2.6 Earth Safety Reference

The earth safety reference is the one true "ground" connection, because the 'ground' conductor in the AC power plug is connected to earth someplace within the building. This connection is intended to help take any AC power line currents that might be present back to the earth connection, rather than through a person who comes in contact with a damaged piece of equipment. The path is a low impedance current path at AC power line frequencies (50/60 Hz) but a high impedance current path at high frequencies, and it cannot be used as a high-frequency reference.

4.3 'Ground' is Not a Current Sink

When we turn on the kitchen faucet, water flows out, and if the drain is open, water flows down the drain hole and into the ground. Although the flowing water analogy is often used to help new students understand electron flow, it does NOT apply in the case of circuit 'ground'. Current does not simply flow into the local 'ground' (or reference) connection and just stay there! As discussed earlier in this chapter, ***all current must flow in a closed loop, and return to its source.***

Signal and power references have been discussed along with how the currents return to their source. When a capacitive filter is used, the unwanted currents do not simply flow into the ground-reference plane and stop. The currents must return to their source. Some path will be found, whether it is the intended path or not. Most emissions problems occur because return currents are using a path that was never intended for the signals associated with these currents.

4.4 Referencing Strategies

There are two basic referencing strategies that are used in PC board design; single-point ground-reference and multi-point ground-reference. While a given product might have a stated design goal of one strategy or the other, nearly all high-frequency circuits become multi-point ground-reference circuits regardless of the intention of the designers. Optimum design requires the engineer to make conscientious decisions about which strategy is possible and desired, given the frequency range of the signals involved in the various circuits.

4.4.1 Single-Point Ground-Reference Strategy

The single-point ground-reference strategy intends to have one point in the system, or on the PC board, where "ground" exists. All circuits are referenced to that point. Figure 4-12 shows a simple diagram of a single-point ground-reference strategy. This works well for DC and low-frequency circuits. Once the frequency is above 100 KHz, however, parasitic capacitance and parasitic inductance become significant enough to allow return currents to flow in unintended paths. Figure 4-13 shows an example of parasitic components added to Figure 4-12. When return currents flow in unintended paths, the likelihood of EMC problems is increased significantly. As stated earlier, most EMI emissions can be traced to return currents flowing where they were never intended.

Single-point ground-reference strategies thus become multi-point ground-reference strategies at frequencies above about 100 KHz. Rather than try to enforce a single-point ground-reference strategy, it is better to consider the circuits separately, and provide a reference path appropriate for that circuit. This requires thought with regard to the return current for each signal, including DC power as well as the low-frequency, middle-frequency, and high-frequency circuits.

4.4.2 Multi-Point Ground-Reference Strategy

As stated in the previous section, regardless of the designer's intention, all referencing strategies become multi-point as the frequency increases beyond approximately 100 KHz. Single-point ground-referencing can still be used for low-frequency and DC power circuits, but engineers should design with a multi-point ground-reference strategy in mind for all signals above approximately 100 KHz.

The "Ground" Myth 57

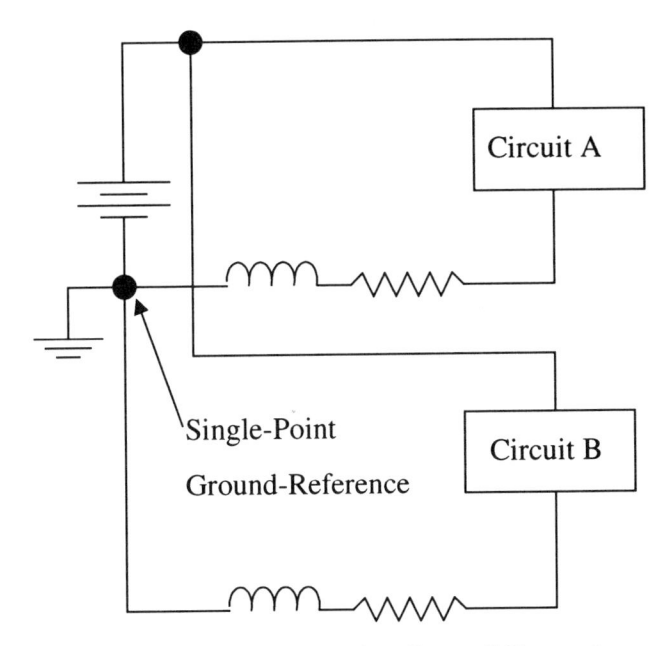

Figure 4-12 --- Single Point Ground Example

A multi-point ground-reference simply implies that each circuit has its own reference. This really makes the most sense, since the return currents must flow back to their source anyway. Indeed, the whole idea of "ground" tends to ignore the return current. When the design is done with the return currents in mind, the concept of "ground" becomes unnecessary. Figure 4-14 shows an example of a multi-point ground-reference strategy. Note that the physical configuration cannot be ignored, nor the circuits drawn as non-physical schematics, because at some location, all the reference points are connected together. To illustrate the multi-point ground-reference strategy, two high-speed circuits are shown with their signal lines and DC power connections. An additional dashed line is shown near the signal lines to illustrate the return path for the high-speed signals. This path is most likely the DC ground-reference plane adjacent to the signal line, but it *could* be a different plane.

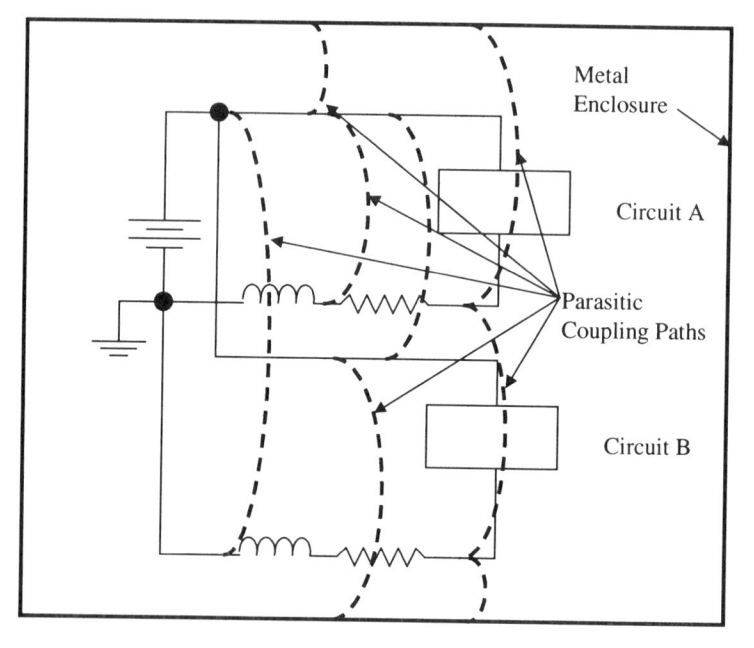

Figure 4-13 --- Single Point Ground Example with Parasitic Coupling

4.5 Grounding Heatsinks to PC boards

It is common to use metal heatsinks placed on top of high-speed ICs to help cool the IC. These metal heatsinks are usually very closely mounted to the high-frequency part of the IC, so internal currents can easily be closely coupled onto the metal heatsink. The metal heatsink is physically and electrically much larger than the IC silicon chip and internal bond wires, so it is a more efficient radiator. Regardless of how the signals are routed on the PC board, or how the return currents are controlled, once the internal IC currents are parasitically coupled onto the heatsink (antenna), they will cause emissions. Depending on the amount of shielding provided (or not provided) by the enclosure, these emissions may exceed the limits.

A common practice for controlling emissions from heatsinks is to "ground" the heatsink to the PC board ground-reference. This will reduce the voltage difference between the heatsink and the ground-reference plane, and reduce the emissions. The number of connections, their size, and their location can make the effort and cost of this heatsink "grounding" effective or may actually increase the emissions [4.2].

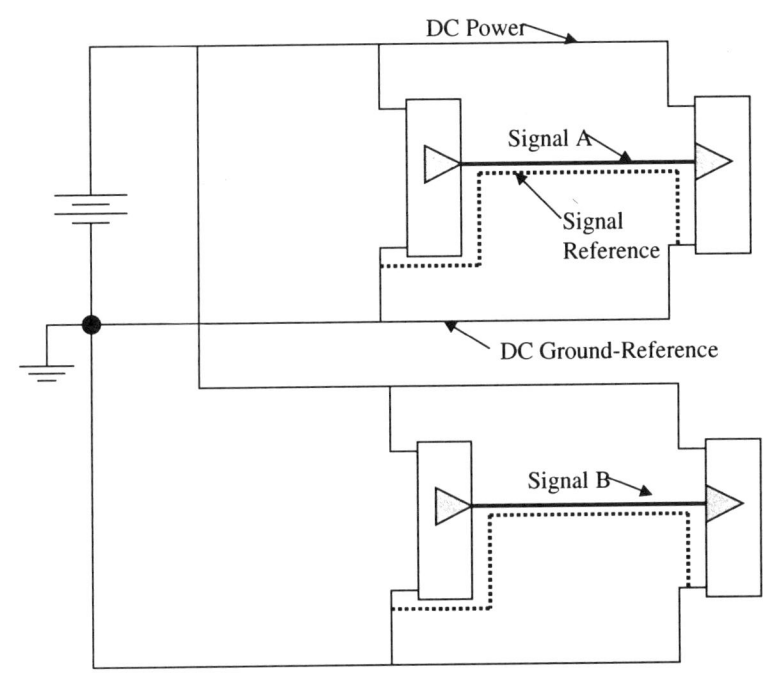

Figure 4-14 --- Realistic Referencing Example

The frequency content of these signals depends on what signals are within the IC. Figure 4-15 shows an example of the voltage between a processor IC and the ground-reference plane measured with a spectrum analyzer. Note that the processor internal clock frequency (1 GHz in this example) was not the dominant frequency detected on the heatsink. The various bus signals contained more current than the internal clock signals and dominated the noise on the heatsink well into the GHz range. As the frequency increases, the size of the heatsink becomes electrically large and makes the heatsink a more effective radiator. Any heatsink "grounding" must therefore be designed carefully in order to be effective at high frequencies.

Heatsink Voltage Measurements for Williamette @ 1000 MHz with 100 Mhz Frontside Bus and 400 MHz RAMBUS

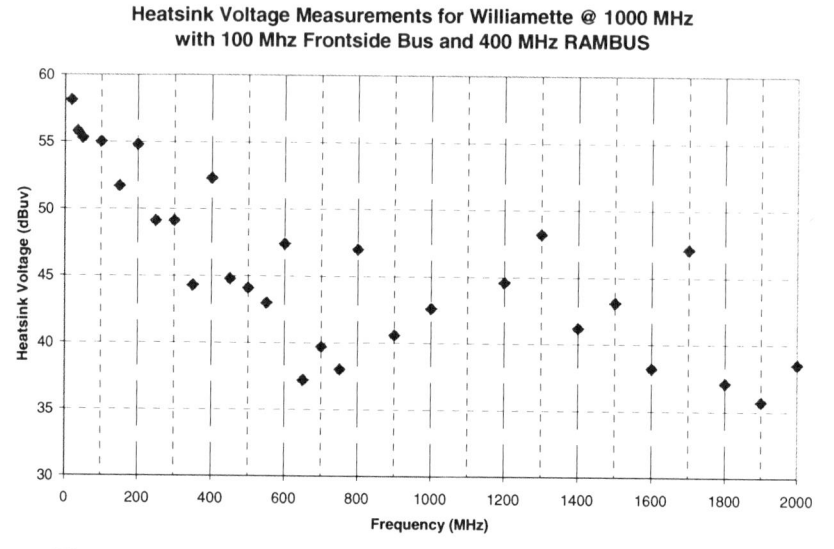

Figure 4-15 --- Example of a Heatsink Voltage Measurement from a Processor IC

As discussed in an earlier chapter, all conductors have inductance. The contacts used between the heatsink and the ground-reference plane will have inductance, and thus will have a non-zero impedance. The greater the number of contacts that are used, the lower the impedance and the more effective the contacts will be in reducing the emissions from the heatsink. Typically, however, the area around the heatsink is valuable real estate and adding "grounding" contact pads is difficult. A trade-off must be made between the number of contacts and the amount of reduction in emissions. The following example is given to show that the emissions can be increased at some frequencies if a sufficient number of contacts are not used.

4.5.1 Heatsink "Grounding" Example

A significant amount of literature has been written about modeling heatsinks and various methods for grounding heatsinks. The Finite-Difference Time-Domain (FDTD) simulation technique is probably the most widely used modeling technique for this type problem, and models have been validated with measurements [4.2], [4.3].

For this example, a 55 mm x 68 mm x 40 mm high heatsink was analyzed. FDTD was used as the simulation technique. An infinite[5] metal plate was used to simulate the metal ground reference plane in the PC board, and the heatsink was placed above the metal plate. The source was then placed between the heatsink and the ground reference plate. Figure 4-16 shows an example of this configuration. Emissions were found in the near field of the heatsink. The highest emission level at each frequency (over a range of positions) was taken as the result, regardless of the direction and polarization.

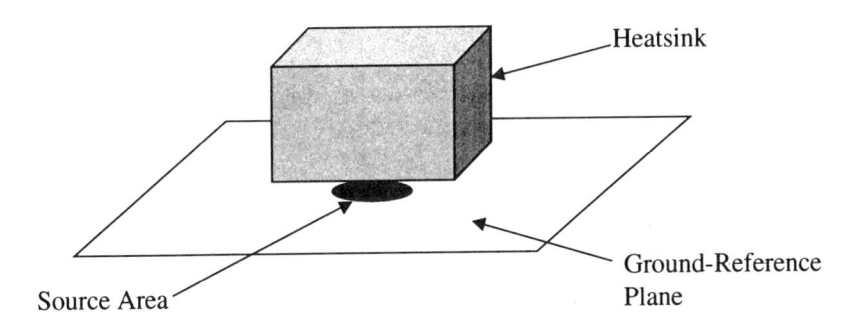

Figure 4-16 --- Heatsink Example Emissions for Few Contacts

A number of different heatsink "grounding" configurations are shown. The no contact case was modeled, as well as one contact point, two contact points (on the opposite sides), four contact points (one in the center of each side), four contact points (one in each corner, and eight contact points (one in each corner, and one in the center of each side). The contact points were small metal posts

---

[5] In FDTD, when the Liao boundary condition is used, a metal plate can be simulated as infinite by placing it against the boundary condition.

62 / PCB Design For Real-World EMI Control

(approximately 25mm x 25 mm) connecting between the ground plane and the heatsink.

The results are given in electric field strength for a given normalized source level. Since the actual source level will vary with different ICs, the absolute level of the result is not important, but the relative levels for different grounding configurations are important and useful results. The results were normalized to remove any influence of the frequency spectrum of the source. That is, the source level was assumed to be identical for all frequencies.

Figure 4-17 shows the results for the no-contact, one contact and two contact configurations. A clear resonance is apparent at 3.75 GHz.

Figure 4-17 also shows that at lower frequencies (300 – 800 MHz), the emissions are increased when only one contact point is used. At frequencies above 800 MHz, the emission levels are approximately the same as those for the case with no contact points. Two contacts improve the low-frequency emissions (below 800 MHz), but increase the emissions from 800 – 2000 MHz. Emissions at frequencies above 2000 MHz when two grounding points are used are approximately the same as no heatsink "grounding".

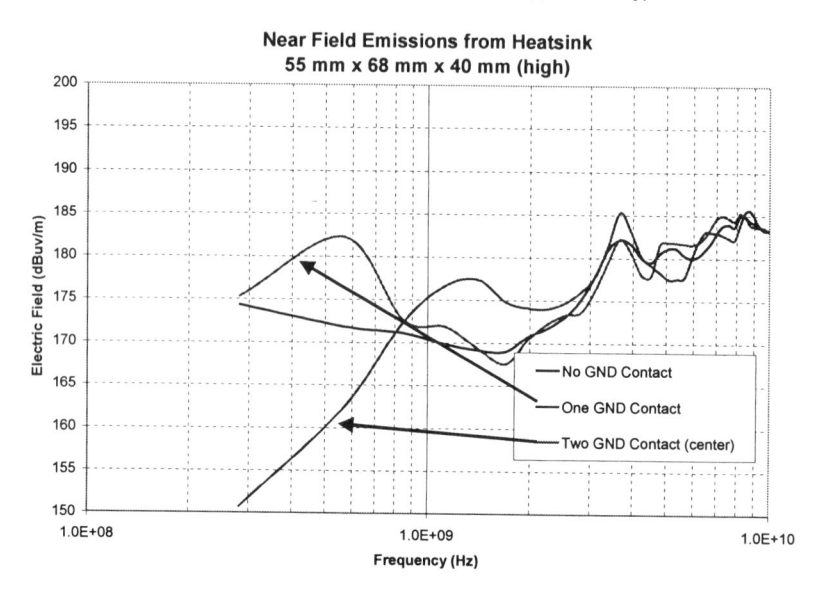

Figure 4-17 Emissions From Heatsink with Few "Grounding" Contacts

Figure 4-18 shows the results when four or eight contacts are used. When the four corners of the heatsink are "grounded" the emissions were reduced below about 1000 MHz. In the frequency range of 1000 – 2000 MHz, the emissions were significantly enhanced. When four contact points were used with one in the center of each side, the emissions were reduced up to about 1600 MHz. Between 1600 and 2500 MHz, the emissions were increased. When eight contact points were used, the emissions levels were significantly reduced up to about 2500 MHz. There was an increase in emissions level between 2500 and 3000 MHz.

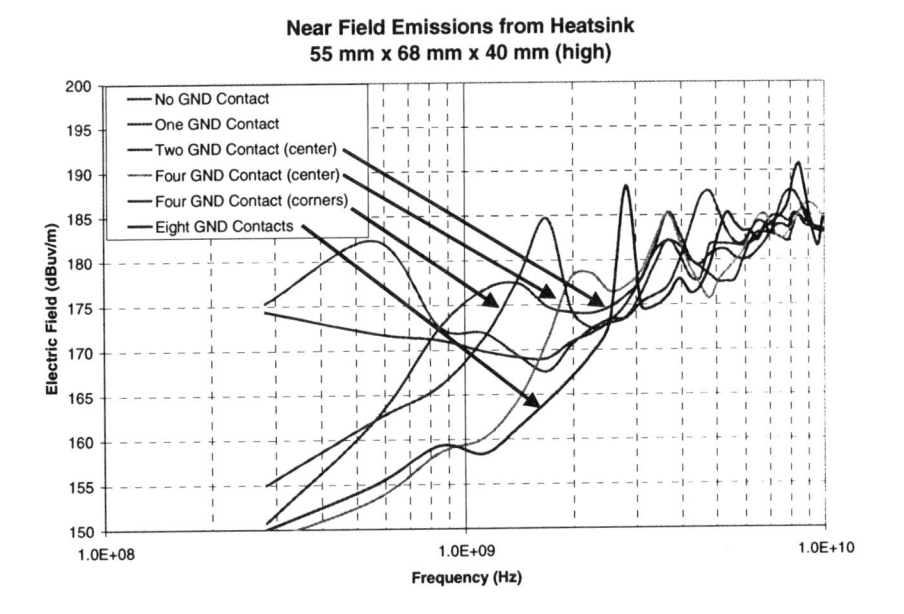

Figure 4-18 Emissions From Heatsink with Additional "Grounding" Contacts

The results show that the emissions levels were greatly affected by the number of contact points between the heatsink and the ground-reference plane. A primary resonance will exist, and depending upon the grounding configuration, the frequency of this primary resonance can be moved to a higher frequency (where it is not going to affect emissions in the frequency range of concern). It is clear from these simulations that one or two contact points should be avoided, since the resonances are likely to be in the range of the first or second harmonic of the processor clock frequency (where the most energy exists). Eight contact points provided the most significant improvement of all the configurations tested.

4.6 PCB Reference Connection to Chassis Reference

If we have a system with a completely enclosed chassis with no I/O cables leaving the enclosure, it makes no difference how, or if, the internal circuits are connected to the enclosure chassis inside the enclosure. Any internal fields from the PC board are contained within the enclosure. This is seldom the case, however, since most real world products include openings and I/O cables.

The most common cause of external emissions in typical products is unwanted common-mode voltages from unintentional signals on the I/O cables and wires relative to the chassis. From an external point of view, we can consider the enclosure chassis to be the reference, and the voltage on the I/O cable to be the feed for an 'antenna'. We might even consider the enclosure and wire (with the source between them) to be a kind of lumpy, irregularly-shaped dipole antenna. Regardless, the combination of I/O wire and enclosure creates an antenna that will radiate quite effectively at some frequencies.

A low impedance connection between the PC board and the chassis, when done correctly, can reduce the emissions from these unintentional signals.

4.6.1 I/O Area Connection

Consider the case with two I/O wires, such as a signal line and its return (ground), (for example, an audio speaker connection to a personal computer). The return wire is usually connected to the ground-reference on the PC board. Any filtering on the audio signal line is referenced to the ground-reference on the PC board. The reference for the external radiation is the chassis and not the ground-reference on the PC board. Any impedance between the chassis and the PC board ground-reference will allow a voltage to be created between them. This voltage will appear to be a source on the I/O reference wire and the I/O signal wire. Figure 4-19 illustrates this problem.

The connection between the PC board reference and the chassis is often accomplished with stand offs and machine screws. The placement of these stand offs is mostly arbitrary and seldom selected for best EMI emissions control. Figure 4-19 shows a connection impedance between the PC board reference and the chassis. This impedance consists of the resistance of the connection and the inductance of the connection. If we consider the loop inductance of the connection between the chassis and the PC board, we will want to keep the loop area as small as possible. Figure 4-20 shows the loop area for the side view and a top view. The total three dimensional loop area is important.

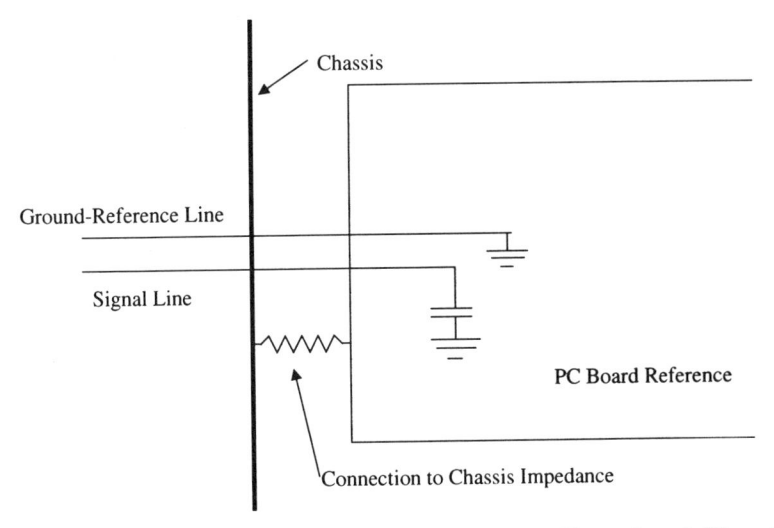

Figure 4-19 Connection Impedance Between Board and Chassis

Many I/O connectors have shields that serve two purposes. First, they shield the connector pins from unwanted signals coupling directly onto them from fields inside the enclosure. Secondly, and more importantly for this discussion, they provide a low impedance and low loop area path from the chassis to the PC board. This

66 / PCB Design For Real-World EMI Control

connector becomes the main connection between the ground-reference plane on the board and the chassis.

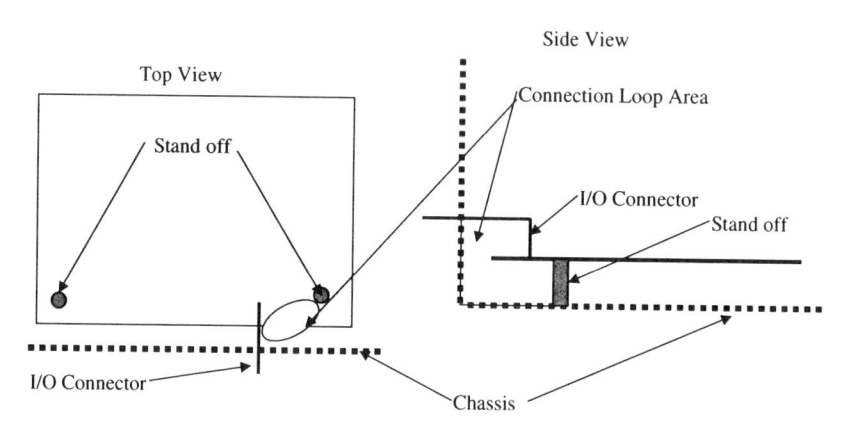

Figure 4-20 Loop Area For PCB Reference Connection to Chassis

4.7 Summary

Unfortunately, it is unlikely that the word "ground" is removable from designer's vocabulary. The concept of "ground" is easy to understand and emotionally comforting. Once the frequency range of the signals is above about 100 KHz, however, the "ground" concept is not a good way to think about the physics of the current flow. Consideration to the return current flow path is vital to controlling "ground" currents.

The term "ground" is often misused to mean a number of different references. It is much better to consider the power-reference, the signal-reference, the chassis-reference, cable shield-reference, etc. Once these various references are clearly described, then the proper connections can be established, and the return currents controlled.

References

[4.1] C. R. Paul, *Introduction to Electromagnetic Compatibility*, Wiley Inter-Science, New York, 1992.

[4.2] C. Wang, J. L. Drewniak, J. L. Knighten, D. Wang, R. Alexander, and D. M. Hockanson, "Grounding of heatpipe/heatspreader and heatsink structures for EMI mitigation," *IEEE Electromagnetic Compatibility Symposium Proceedings*, Montreal, Canada, pp. 916-920, August 2001.

[4.3] K. Li, C. F. Lee, S. Y. Poh, R. T. Shin, and J. A. Kong, "Application of FDTD method to analysis of electromagnetic radiation from VLSI heatsink configurations," *IEEE Trans. EMC*, vol. 35, pp. 204-214, May 1993.

Chapter 5

Return Current Design

5.1 Introduction

Many EMI problems associated with high-speed circuits are due to improper design of the return current path. PCB designers often expend considerable effort to carefully design traces with the proper length, proper transmission line impedance, etc., but neglect the return current path that completes the current loop. These EMI problems can usually be avoided if attention is paid to this return current path.

First, to understand the physics of the high-speed transmission line, the concept of current flow must be extended beyond the direct current (DC) sense. Current does not start at the source, travel down the transmission line signal conductor, appear at the receiver, and then return through the ground-reference plane back to the source. The transmission line supports a transverse electromagnetic (TEM) wave. [5.1] When the transmission line is longer than the pulse, there will be regions along the transmission line where the TEM pulse is present, and regions where the TEM pulse has not yet arrived (or has already passed).

In the region where the TEM pulse exists, there will be a time-varying electric field between the trace and the ground-reference plane, as shown in Figure 5-1a and Figure 5-1b for the simple microstrip and stripline cases, respectively. The electric field requires a current to be in the two conductors. This current will be in opposite directions in the conductors, and must exist simultaneously with the pulse. The current must flow in both conductors as the pulse moves down the trace. Any discontinuity or break in the current's path in either the trace or the ground-reference plane will affect the current. The current must flow in order for the TEM pulse

70 / PCB Design For Real-World EMI Control

to propagate, and it will likely find a path that compromises the signal integrity or EMI performance, if the designer does not provide an intentional path.

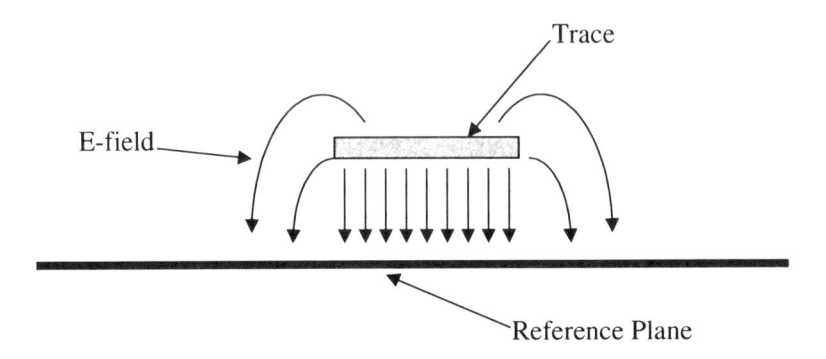

Figure 5-1a Electric Field for Microstrip

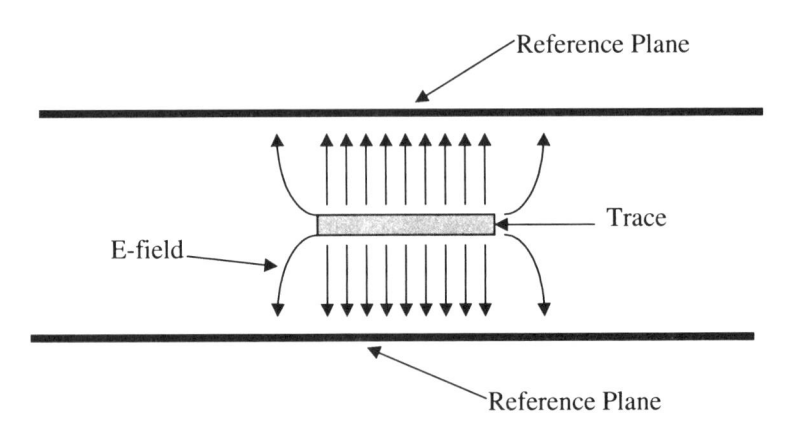

Figure 5-1b Electric Field for Stripline

There are three different ways that the return current path is often interrupted: a split in the reference plane, a signal trace changing reference planes, or a signal going through a connector between two different circuit boards. Each of these three design concerns will be addressed separately.

5.2 Split Reference Planes

It is important to remember that the return current is present due to the TEM pulse traveling down the transmission line. This signal return current is not affected by the DC power schematic name given to the reference plane. That is, the return current is the same whether the plane is called a "ground" plane, or a "VCC" plane, or whatever. "Ground" planes and power planes are both used as reference planes for signal return currents in many high-speed designs. The ground-reference plane is usually unbroken and provides a good return current path. The power planes, however, are often broken into smaller areas to allow different voltages, or DC power supplies, to exist on the same PCB layer. When a trace is run across a split between two different DC power voltages, etc. the return current can not flow across the split. Figure 5-2 shows an example of a trace crossing a split. Regardless of whether the split is in the power plane or a "ground" plane, a high-speed trace should not cross a split in its reference plane. Note that when the power plane is used as the reference plane, the current must still return to the driving IC's ground-reference pin. The current must flow through decoupling capacitors to flow from the power plane to the ground-reference plane.

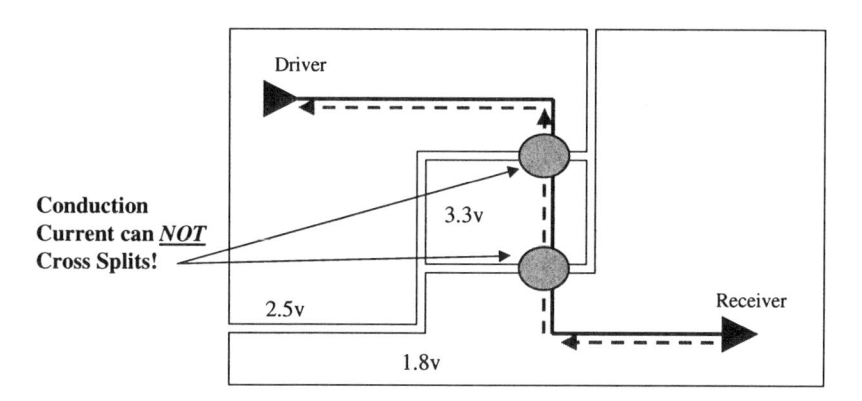

Figure 5-2 Example of Trace Crossing Split Reference Planes

5.2.1 Stitching Capacitors

While it is clear that high-speed traces should not be run across a split in the reference plane, design constraints sometimes make this necessary. For example, when a six layer PCB is being used, the most common layer assignments are four signal layers and two plane layers in a S-P-S-S-P-S stackup as shown in Figure 5-3. One of the planes is typically assigned as "ground" and is not usually broken. The other plane is usually assigned as the power layer, and all the different voltages required for the different ICs and circuits are positioned in the required areas on power islands. Since there are only four signal layers, and two of them are referencing the power plane layer, there is little chance to completely avoid a high-speed trace crossing a split between two different power islands. For example, in Figure 5-3, if Layer #5 is the power layer, the signal return current from signals on Layer #6 will use the underside of Layer #5 as its path. The signal on Layer #4 will be most tightly coupled to Layer #5 and will return on the upper side of Layer #5. Due to skin effect, these currents will remain on separate sides of Layer #5. Stitching capacitors connected between two power islands and positioned close to the location where the trace crosses the split are often used to try to provide a return current path across the split. Figure 5-4 shows an example of stitching capacitors across splits in the reference plane.

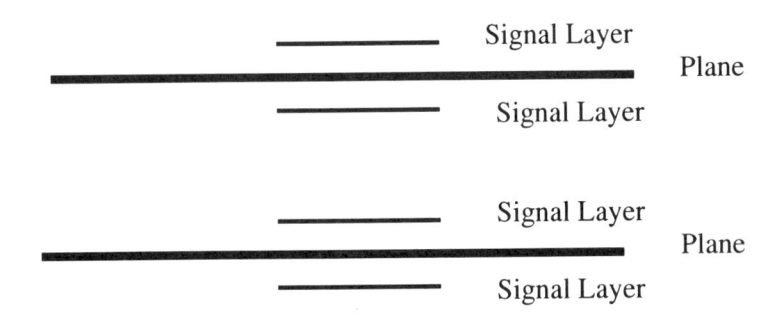

Figure 5-3 Six-Layer PCB Stackup

Return Current Design / 73

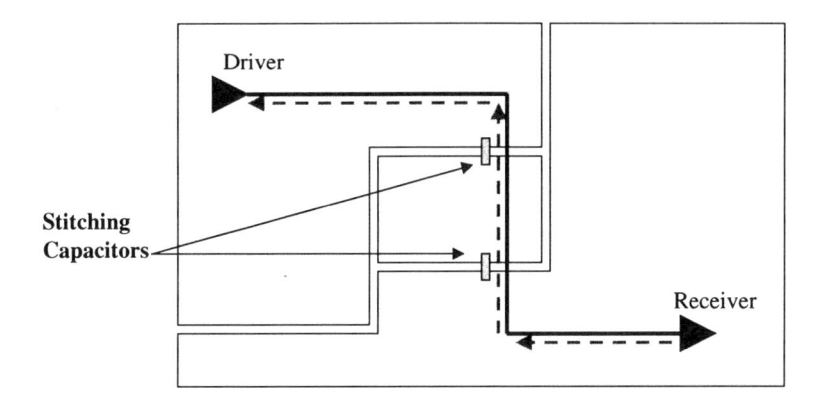

Figure 5-4 Stitching Capacitors Across Split in Reference Plane

While the use of stitching capacitors helps provide a return current path across a split, the impedance of that path must be considered. Certainly, if no split exists, the impedance along the plane is quite low. The stitching capacitor must provide that low impedance to allow the return current to cross the split. Recall, however, that the impedance of a capacitor changes with frequency. Figure 5-5 shows a typical surface mount capacitor's measured impedance as a function of frequency. This capacitor is a 0.01 uF capacitor in an 0805 size SMT package. As the frequency increases, the capacitor's series impedance decreases until the self resonant frequency is reached. At the self resonant frequency, the inductance of the capacitor becomes dominant, and the impedance increases with frequency. If a 100 MHz clock signal is the intentional signal on a trace crossing a split, the magnitude of the stitching capacitor's impedance is approximately 0.4 ohms at 100 MHz. In order to support a fast rise time on the clock signal, harmonics of the fundamental frequency are also required. The 9th harmonic (at 900 MHz) sees a much higher impedance, approximately 4-5 ohms. The return current at the fundamental frequency would have a much lower impedance path through the stitching capacitor than the would this signal's higher harmonics.

Figure 5-6 shows an example of the radiated field from a simple PCB with an exposed microstrip trace. Both the non-split plane and split plane (without a stitching capacitor) cases are shown over a

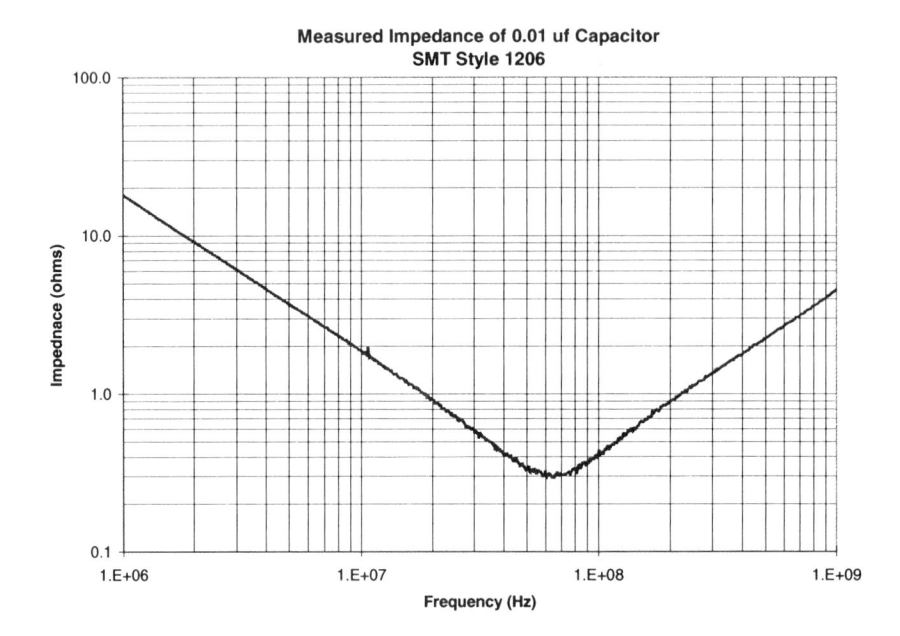

Figure 5-5 Typical 0805 SMT 0.01 uF Capacitor Impedance

frequency range of 20 – 1000 MHz. The radiated fields are much greater, in excess of 20 dB, when the trace is crossing a split, as expected. Figure 5-7 shows the same example when one or two 0.01 uF stitching capacitors are placed across the split. For frequencies up to approximately 100 MHz, the emissions are reduced to the same level as the case when there is no split. When only one stitching capacitor is used, the emissions above 100 MHz increase as the frequency increases. When two stitching capacitors are used, the emissions remain low up to 1 GHz.

Until now, an ideal capacitor as the stitching capacitor has been discussed. The inductance associated only with the current path through the capacitor body was used, but the mounting and connection metal on the PCB has been ignored. If the solder pads, via to the plane, and connecting traces are considered, an additional 1.5 nH of inductance must be included for the two connections. This will result in much greater impedance at high frequencies, and would be expected to increase the emissions. Figure 5-8 shows the

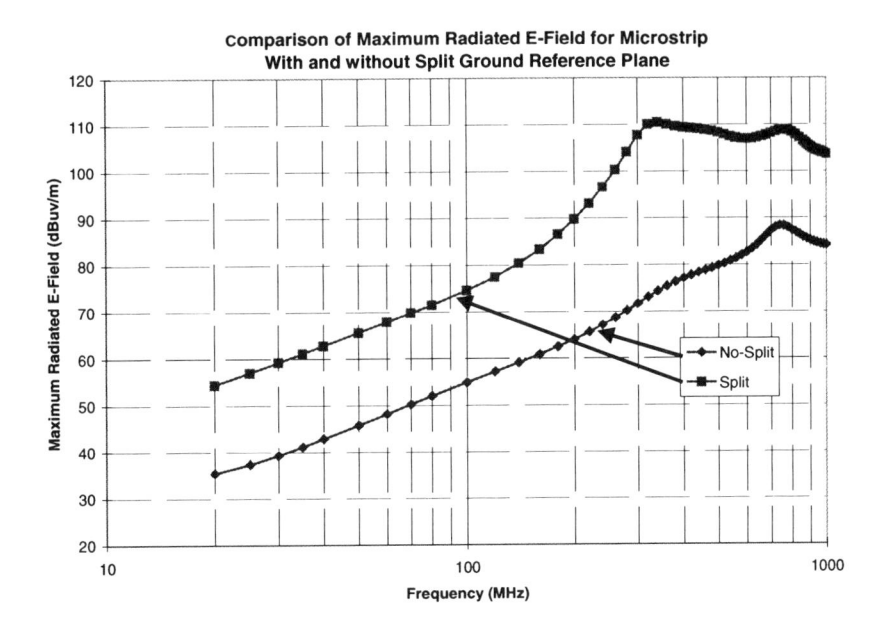

**Comparison of Maximum Radiated E-Field for Microstrip
With and without Split Ground Reference Plane**

Figure 5-6 Near Field Radiation from a Microstrip on a Board with a Split in the Reference Plane

radiated emissions from the same example when the inductance of the connection vias etc. are included. The emissions are higher when the capacitor is more accurately represented than when the additional inductance is ignored (and again significantly greater than the case with no split), but are still lower than in the case where no stitching capacitors bridge the split.

The best approach is to not cross a split with high-speed traces. If a PCB stackup requires splits in a power plane, then careful layout of the high-speed traces so that they are referenced only to a continuous (or solid) "ground" plane, or to areas where the power planes are not split, will provide the best EMC design.

76 / PCB Design For Real-World EMI Control

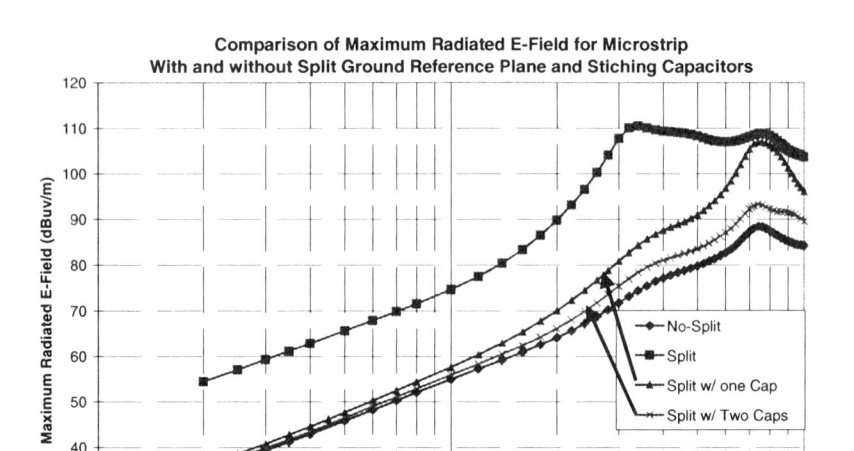

Figure 5-7 Near Field Radiation from a Microstrip on a Board with 'Perfect' Stitching Capacitors

5.3 Trace Changing Reference Planes

It is very common to route a high-speed trace on multiple layers. In order to use as many routing channels as possible, designers will usually run traces on one layer in a north-south orientation, and east-west orientation on another layer. This means the signal must change layers to reach the receiver, and the return current must change reference layers as well. Figure 5-9 shows a simple example of a trace changing layers through a via. This example shows a basic four-layer board, with the signal layers on the outside (top and bottom). A simple diagram of the return current is shown in Figure 5-10. How can the return current cross from the bottom reference layer to the top reference layer? There are two parallel paths. For lower frequencies, the path is through a decoupling capacitor positioned nearby. For higher frequencies, the path is through the displacement current of the interplane capacitance. The path with

Return Current Design / 77

the most current will depend on which path has the lower impedance at a given frequency.

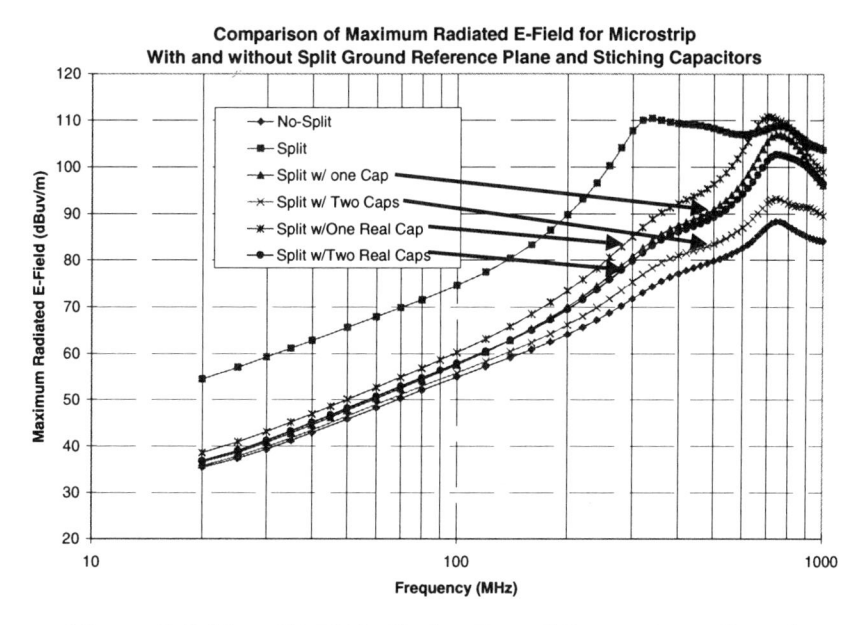

Figure 5-8 Near Field Radiation from Microstrip on Board with Stitching Capacitors Including PCB Connection Inductance

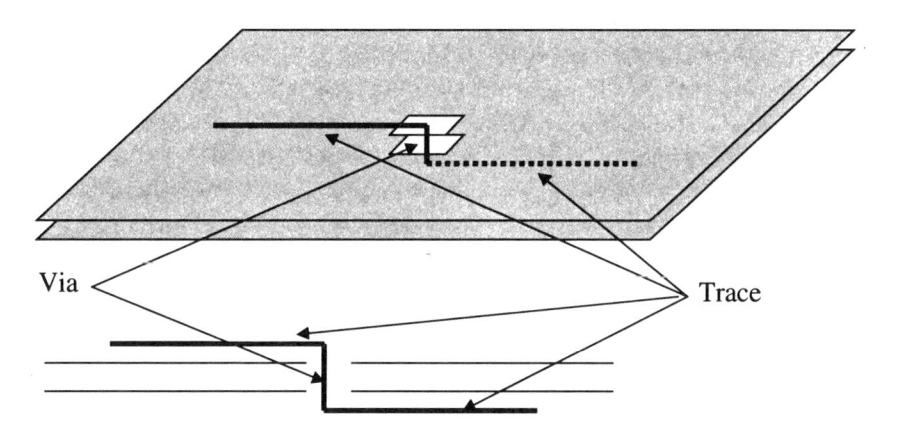

Figure 5-9 Traces Through a Via and Changing Reference Planes

78 / PCB Design For Real-World EMI Control

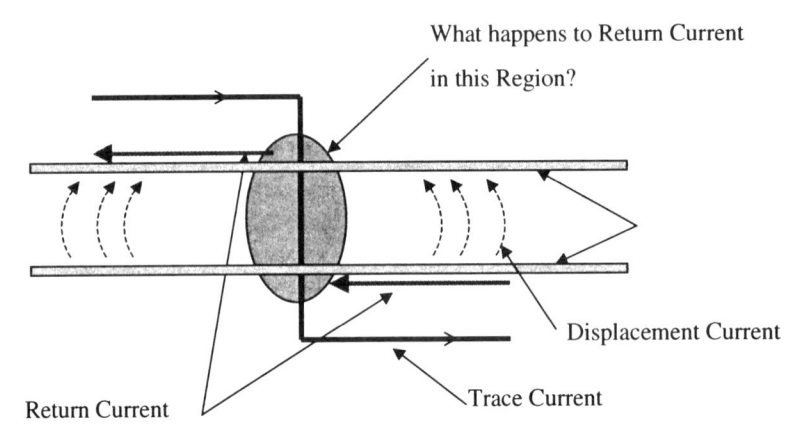

Figure 5-10 Return Current Near Via

Often, designers place decoupling capacitors near the via transition to facilitate the return current path. Figure 5-11 shows the return current path through a decoupling capacitor. The current cannot flow through the plane, due to skin depth, so it must flow around the via hole opening. The current will be on the underside of the bottom plane, flow to the via hole opening, travel on the bottom surface of the bottom plane to the capacitor, through the capacitor, travel on the inside surface of the top plane to the via hole, through the via hole opening, and finally under the trace on the top surface of the top plane. Naturally, the impedance of the capacitor and the inductance of the connection traces and vias will affect the return current path as discussed in Section 5.2.1.

Once again, this is an greatly simplified view. Decoupling capacitors are not placed between the planes, but rather are placed on the surface layers of the PC board. Figure 5-12 is a better representation of the current paths through decoupling capacitors. Again, the currents flow to the decoupling capacitor, but now, some of the currents will become common-mode currents and will flow on the outside of the top plane where they can cause direct emissions.

Naturally, when the impedance through the decoupling capacitor is higher than the displacement current path between planes, there will be little advantage using the decoupling capacitor. When the displacement current path is the lower impedance path, these currents cause 'noise' between the planes over a widely distributed area.

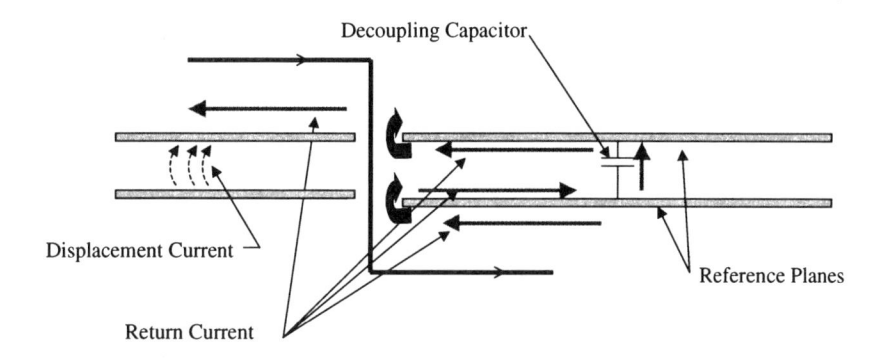

Figure 5-11 Return Current Through a Decoupling Capacitor

If the reference plane must be changed, then decoupling capacitors should be used close to the via where the layers are changed. While this will not help at high frequencies, it will help lower frequencies. As in the case of the stitching capacitors in section 5.2.1, two capacitors placed very close to the via should be used to minimize the additional emissions. The capacitors must be selected to have low impedance across the frequency range that includes the harmonics of the fundamental signal on the trace changing layers.

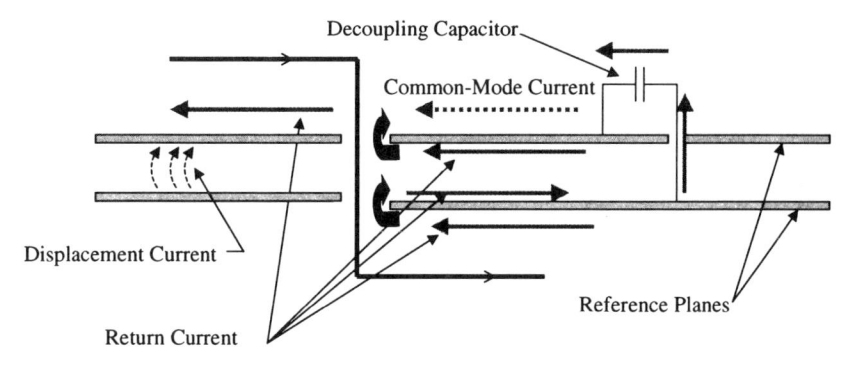

Figure 5-12 Return Current Through Decoupling Capacitor on the Surface of the PC Board

Naturally, it is best not to change reference layers. This does not mean the trace must remain on a single layer in the PCB stackup, but that care must be taken when changing layers. Figure 5-13 shows a simple six-layer board stackup. Figures 5-13a and 5-13b show a routing strategy where the trace changes reference planes, and decoupling capacitors must be used. Figure 5-13c shows a better routing strategy where the trace changes layers but does not change the reference plane (only the surface of the reference plane used for the return current changes). Return currents can flow from one side of the plane to the other side of the plane through the anti-pad opening of the via hole, and no extra currents are present to cause emissions, nor is the impedance of the return current path increased with a decoupling capacitor and its associated interconnect inductance.

5.4 Motherboards and Daughter Cards

It is very common to have two PCBs, such as a motherboard/daughter card arrangement, where high-speed signals must travel from one board to the other. In a simple example of two four-layer boards (two signal layers and two plane layers), the connection between the planes would likely be as shown in Figure 5-14. A signal trace is also shown, and it is referenced to the "ground" plane on the motherboard and the power plane on the daughter card. The return currents are now added in Figure 5-15. Since the assignment of the signal trace is to layers referenced to different planes, this is effectively the same as the case in Section 5.2. The return current will take the path of least impedance, which will be through decoupling capacitors or through the interplane capacitance displacement current, depending upon the frequency. As in the previous section, higher frequencies will tend to use the displacement current path, while lower frequencies will use nearby decoupling capacitors. When the path is through the decoupling capacitors, the return currents must find the nearest decoupling capacitor(s), and may use capacitors from either board or both boards to return to the necessary reference plane. If the return current path is much longer than desired, exposed currents will most likely increase emissions and these currents can lead to functional problems. When the displacement current path is the lower impedance path, these currents cause 'noise' between the planes over a widely distributed area [5.2] and can cause common-mode currents.

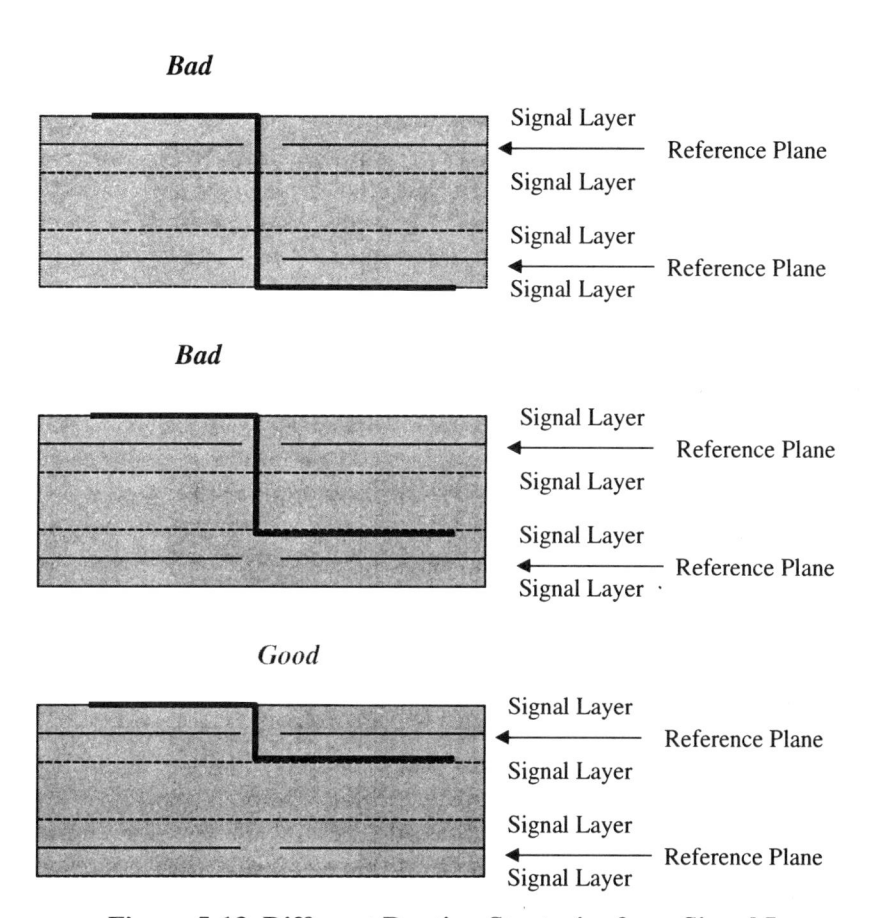

Figure 5-13 Different Routing Strategies for a Signal Layer Change

The solution is to use foresight when designs include card-to-card connectors. Figure 5-16 shows the return currents when the signal is assigned to the same reference plane on both cards. The return currents stay closely coupled to the signal traces and emissions are greatly reduced.

82 / PCB Design For Real-World EMI Control

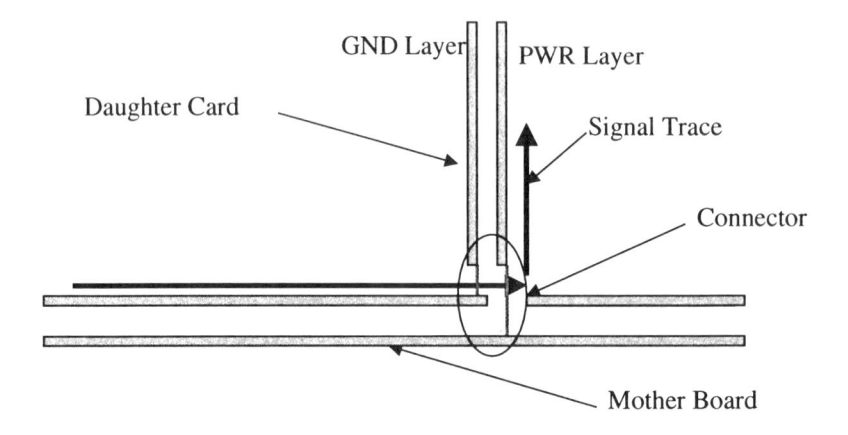

Figure 5-14 Motherboard/Daughter Connection Routing

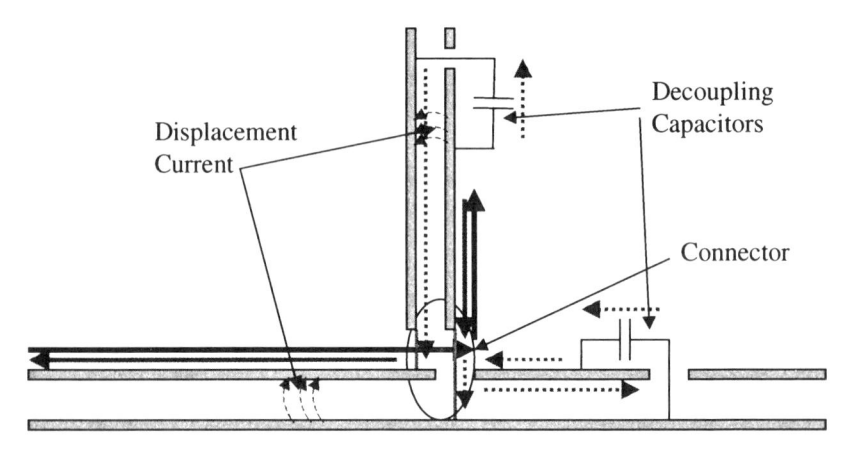

Figure 5-15 Motherboard/Daughter Connection Return Current

5.4.1 Connector Pin Assignments

Connector pin assignments are traditionally a subject of debate among EMC design engineers. Many designers simply try to assign as many "ground" pins as possible, while others insist that alternating signal and ground pins is the best strategy. Both of these approaches ignore the power pins, except to acknowledge that power pins are required and a sufficient number must be placed in the connector to provide a low DC voltage drop across the connector.

Both of these approaches ignore the need to manage the return current path to optimize the emissions from the connector. As stated earlier in this chapter, the schematic name of "ground" or "power" is not important to the return current. *The best design strategy is to match a signal line with a return path, regardless of its schematic name.* If a 64 bit bus is routed through a connector, and 40 of the signals are referencing the power plane, and only 24 of the signals are referencing the "ground" plane, then the best strategy is to use 40 power pins and only 24 "ground" pins, and to place these pins as close to the signals using them as a signal return as possible.

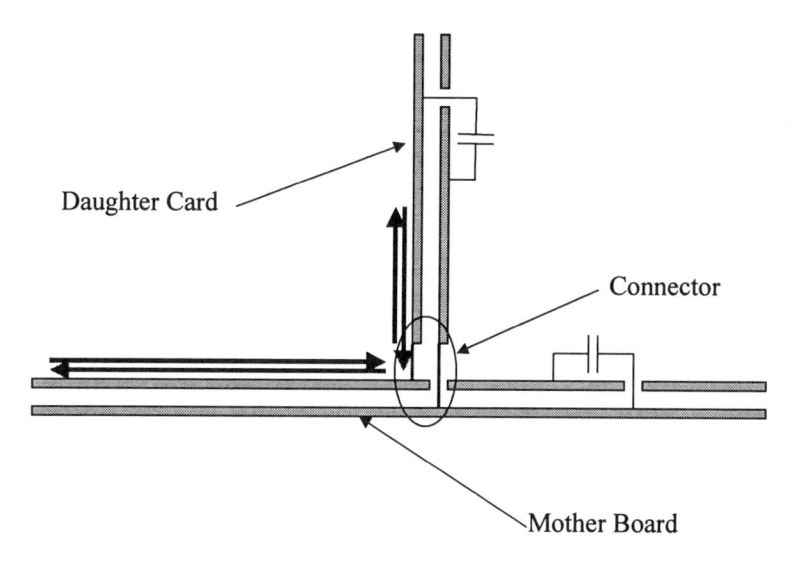

Figure 5-16 Optimum Motherboard/Daughter Card Connection Return Current

5.5 Summary

The return currents associated with high-speed signal traces are critical to the EMI design success. The path these return currents will take must be intentionally designed or the return currents will most likely take a path that causes additional emissions and may result in functional problems.

The most common return current path problems are due to splits in the reference plane, changing reference plane layers, and signals through connectors. Stitching capacitors or decoupling capacitors may help in some cases, but the impedance of the capacitors *and* the PCB connection vias, pads, and traces must be considered for the fundamental frequency and the harmonics needed to support the fast rise times.

Developing a design strategy that considers the return current path before the traces are actually routed on the board will provide the best chance for success in lowering radiated EMI emissions. Since changing the routing layer etc. before the initial layout does not cost money, this is a way to improve EMC for free!

References

[5.1] D. M. Pozar, *Microwave Engineering, 2nd Ed.*, John Wiley, New York, 1998.

[5.2] W. Cui, X. Ye, B. Archambeault, D. White, M. Li, and J. L. Drewniak, "Modeling EMI resulting from a signal via transition through power/ground layers," *Proceedings of the 16th Annual Review of Progress in Applied Computational Electromagnetics*, Monterey, CA, pp. 436-443, March 2000.

Chapter 6

Controlling EMI Sources
– Intentional Signals

6.1 Introduction

When starting to think about EMI design, the most effective approach is to consider the actual sources of EMI emissions, and treat them all individually. Most of the causes of EMI emissions at the printed circuit (PC) board level are separable and can be considered and treated individually, without increasing the emissions from some other source.

In order to do this, all the signals on the board must first be separated into two categories: intentional signals and unintentional signals. During the board design, engineers naturally consider the intentional signals. These are the signals they 'intend' to be on the board, and carefully design traces on the board to carry them from their source to their destination.

Unintentional signals are often neglected during the design. After all, we don't 'intend' them to be on the board, so they are not often in our thoughts. These unintentional signals are the cause of more than 90% of the EMI emissions from a PC board! Some of these unintentional signals will be present regardless of how carefully the board is designed. They must be considered and appropriate steps taken to insure they will not cause excessive emissions.

The sources of emissions can therefore be broken into two major categories: intentional signals and unintentional signals. Each of these major categories can be further broken down. Sources of emissions from intentional signals include loop-mode[1] and common-mode[2] sources. Sources of emissions from unintentional signals include common-mode, crosstalk coupling to I/O traces, power planes, and above board structures. This chapter will describe the sources (and how to control them) resulting from intentional signals and Chapter 7 will describe the sources resulting from unintentional signals.

6.2 Critical Signals

Naturally, not all signals on a PC board cause EMI emissions that concern us. Many signals only switch occasionally or only on system start up. It is necessary to focus only on those signals that switch often. Clock signals, memory, data, and address bus signals, data strobe signals, video signals, and any other fast, high bandwidth signals must be considered EMI critical signals. These signals must be all carefully considered, since they all are possible sources of EMI emissions

6.3 Intentional Signals

The intentional signal is usually specified as a certain data rate and a rise/fall time. Since EMI emissions limits are specified in the frequency domain, the time domain intentional signal must be converted into a frequency spectrum representation. A general pulse waveform contains a number of different sine waves of various amplitudes and phases. For example, a typical clock signal is a simple squarewave. A squarewave can be constructed from a sine wave of the fundamental frequency, and all its odd numbered harmonics, each in-phase and with a decreasing amplitude. Figure 6-1 shows a simple in-phase summation of the fundamental, third, fifth,

---

[1] The term "differential-mode" has different meanings to different people. We will define "loop-mode" in Section 6.4.
[2] The term "common-mode" also has different meanings to different people. We will define this term in Section 6.6.

seventh, and ninth harmonics. With only these few harmonics included, the sum is already looking like a squarewave with some ripple.

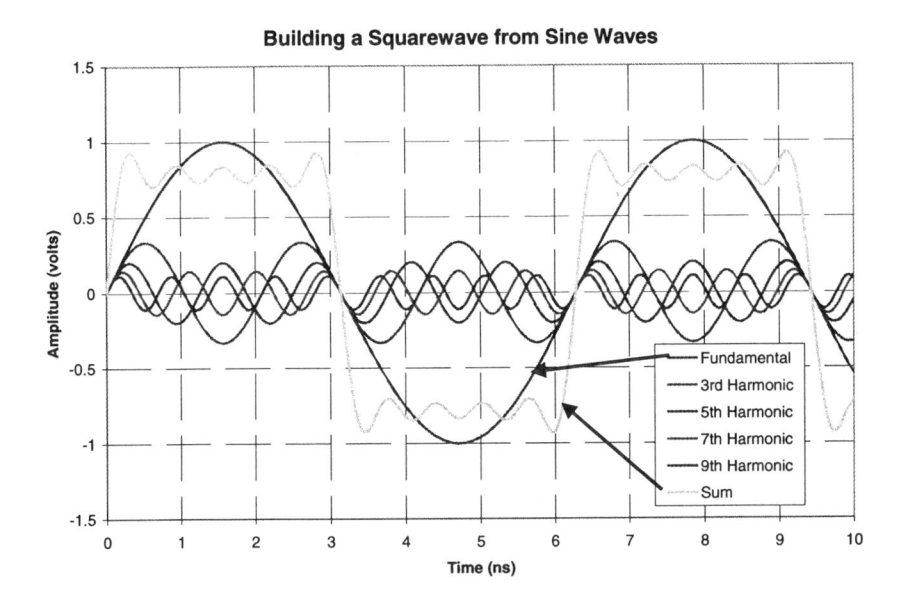

Figure 6-1 Harmonic Content of a Squarewave

Figure 6-2 shows the envelope of the frequency spectrum of a typical trapezoidal pulse based on the pulse width and the rise and fall time of the pulse. Since higher frequencies tend to radiate more efficiently from traces, etc., and also radiate through smaller openings in the metal enclosure [6.1], [6.2], it is best to keep the amplitude of the high-frequency harmonics as low as possible. As Figure 6-2 shows, the amplitude of the pulse spectrum decreases with increasing frequency. The spectrum for a given voltage amplitude pulse decreases at a rate of 20 dB per decade of frequency above a frequency related to the pulse width, and at 40 dB per decade above a frequency related to the pulse rise/fall time. The slower the rise/fall time for a given pulse amplitude, the lower the second break frequency becomes, resulting in reduced signal levels at high frequencies. Clearly, the slower the rise and fall times of a pulse, the lower the potential frequency domain harmonic content of that signal.

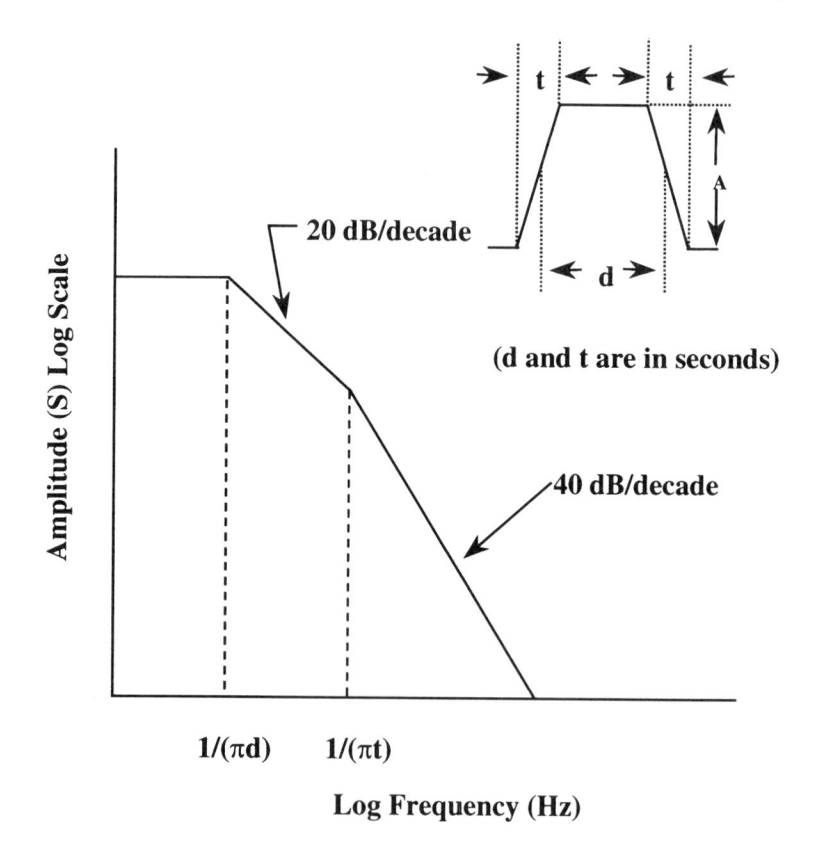

Figure 6-2 Spectrum Envelope for Trapezoidal Pulse

While "rise" and "fall" times are commonly used to discuss effects on the frequency spectrum of a signal, consideration of the edge rates are critical for comparing signals with different pulse amplitudes. Rise time and fall time are normally defined as the time it takes for the pulse to rise (or fall) between the 10% and 90% amplitude levels of the pulse. Consider a 5 volt pulse with a 1 nsec rise time, the edge rate is 5V/ns. If the pulse is changed to a 2.5 volt pulse with the same edge rate, the rise time is now only 500 psec. The shorter rise time would normally cause significant concern, but in this case it does not add to the high frequency spectrum content because the edge rate is constant. Lowering the pulse amplitude would actually lower the overall spectrum without changing the knee frequencies in Figure 6-2. An example of the edge rate effect is shown in Figure 6-3. In general, when we discuss rise and fall times, and their effect of the signal spectrum, we are keeping the pulse amplitude constant.

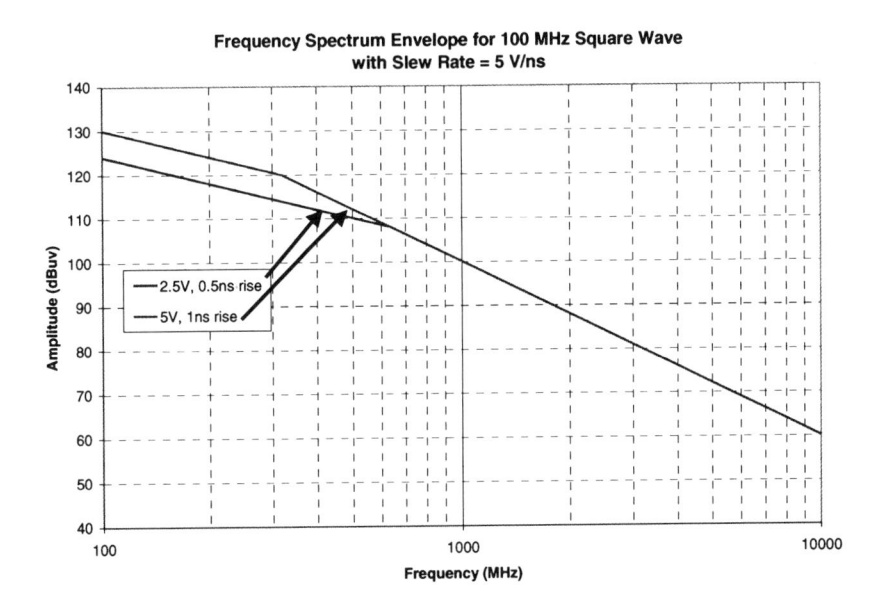

Figure 6-3 Comparison of Frequency Spectrum for Different Pulse Rise Times and Amplitudes

Figure 6-4 shows the relative frequency spectrum amplitude for different data rates and rise times (for the same pulse amplitude). The change of the data rate makes almost no difference in the high-frequency harmonic amplitude, but changes in the rise and fall times[3] make a significant difference in the same high-frequency harmonics.

---

[3] Rise and fall times are changed while keeping the pulse amplitude constant.

90 / PCB Design For Real-World EMI Control

Comparison of Various Data Rates and Rise/Fall Times on the Frequency Spectrum

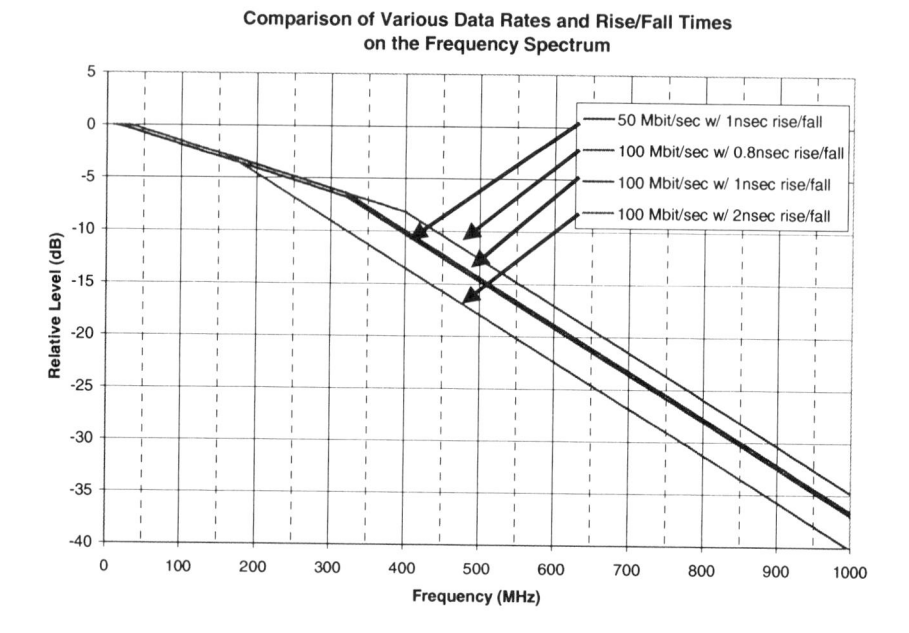

- 50 Mbit/sec w/ 1nsec rise/fall
- 100 Mbit/sec w/ 0.8nsec rise/fall
- 100 Mbit/sec w/ 1nsec rise/fall
- 100 Mbit/sec w/ 2nsec rise/fall

Figure 6-4 Frequency Spectrum for Different Data Rates

While all the calculations of the harmonic content of the frequency spectrum of a given pulse are useful, the actual real-world pulses seldom resemble a clean trapezoidal pulse. Slight imperfections in the waveform can have dramatic effects on the high-frequency harmonic amplitude. Figure 6-5 shows an example of two different voltage waveforms, and Figure 6-6 shows the associated frequency spectrum for both voltage waveforms.

Typically, the signal integrity analysis of critical signals uses the voltage waveforms, and the example waveforms in Figure 6-5 are nothing extraordinary. In fact, for most applications, either waveshape would be acceptable, since the transition region is single directional (either rising of falling) and double switching would not occur. The additional noise on the top or bottom of the waveshape would often be within the voltage noise margins for a given signal integrity analysis.

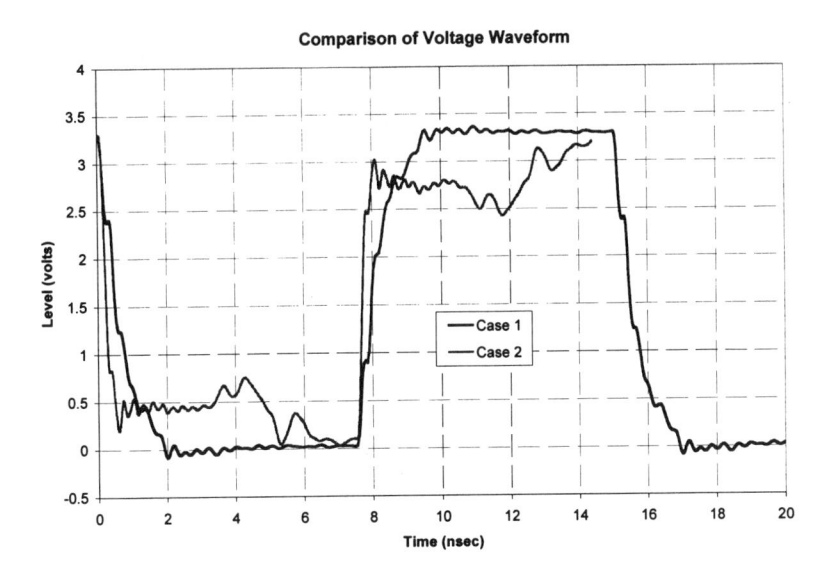

Figure 6-5 Voltage Waveform (Time Domain)

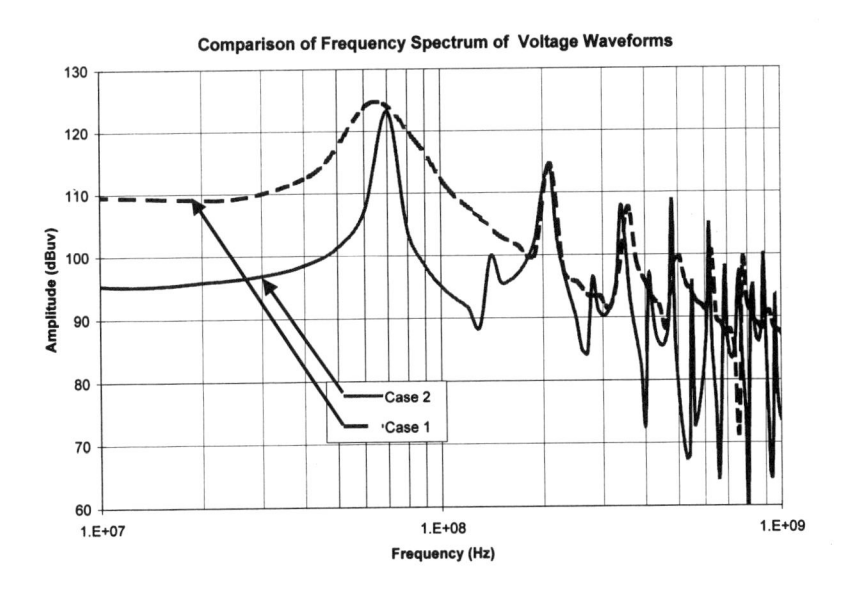

Figure 6-6 Voltage Waveform (Frequency Domain)

For EMI applications, however, the current is the most important consideration. Current radiates, not voltage! Equation 6.1 gives the relationship to find the electric field from a current loop in free space [6.3]. Since most of this equation would be constant for a given frequency and observation point, Equation 6.1 reduces to a much simpler expression in Equation 6.2

$$E_\theta = \frac{\omega \mu IS}{4\pi} \left(\frac{\beta}{r} + \frac{1}{jr^2} \right) e^{-j\beta r} \sin\theta \qquad (6.1)$$

$$E_\theta = kIS \qquad (6.2)$$

where
k = a constant for a given frequency and observation point,
I = the amplitude of the current, and
S = the area of the conductor loop.

It is clear that the radiated electric field is dependant on only two things: the magnitude of current and the size of the loop.

From this example, it can be seen that the current waveform on a critical net is very important for EMI purposes. With the current IC technology, the current and voltage are not the same waveshape, as would be the case in a simple resistive circuit. The signal integrity analysis of critical signals should include the analysis of the current waveforms for EMI purposes, once the signal integrity voltage waveform analysis is complete. Figure 6-7 shows the time domain current waveforms for the critical signal traces whose voltage waveforms were just discussed, and Figure 6-8 shows the frequency spectrum for the currents of the two signals. In this figure, it is apparent that the high-frequency content of the current is very different for the two signals, and it is expected that their emissions will be very different as well.

6.4 Intentional Signals – Loop-Mode

Recall that we are considering the various possible sources of emissions separately. In this section, we will discuss loop-mode

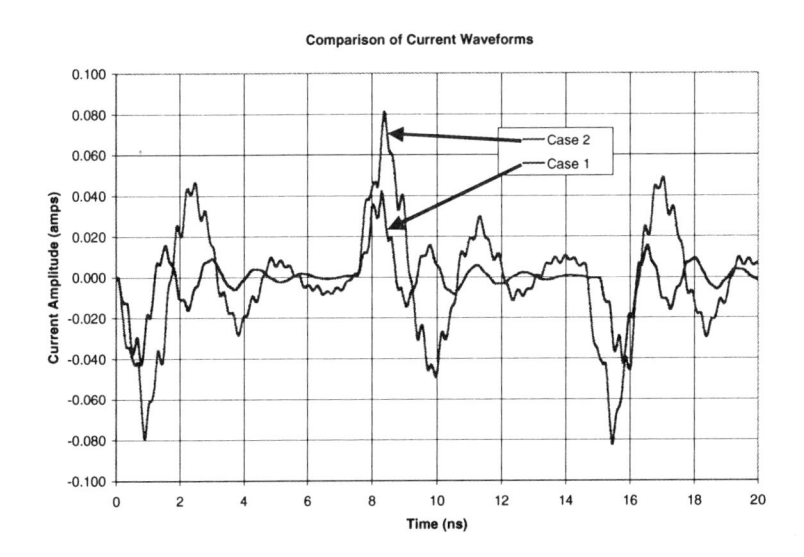

Figure 6-7 Current Waveform (Time Domain)

emissions from intentional signals. Direct emissions from intentional signal traces are considered to be loop-mode emissions. That is, the current flows down a trace, and returns under that trace in the reference[4] plane when a microstrip is used on a multilayer board. If a board without reference planes is used, then the current return must be through an explicitly-designed return trace. For most high-speed signals, the microstrip, stripline, or asymmetric stripline PCB stackup is used, and the emissions from these are defined by the amount of current exposed on the outside layer of the PCB for the loop-mode potential source.

---

[4] Please refer to Chapter 4 for a discussion about "ground" and "reference" planes for signal return paths.

The current path forms a small loop between the trace and the reference plane below the trace. This can be analyzed using a simple loop antenna approach. While there are commercial software tools that will provide the emissions from this simple loop antenna at any specified distance (such as 10 meters), this analysis is not very useful and often misleading. The analysis should be limited to the near field, often about 1" – 2" above the PCB. Nearly every product with a PCB has some amount of metal shielding around it and/or long wires attached to it. The far field emissions will usually be dominated by signals on the long wires and not signals radiating directly from the PCB traces. Certainly, any amount of metal shielding around the product will completely change the nature of the emissions from the traces themselves, so analyzing the simplified loop is hardly representative of the actual design.

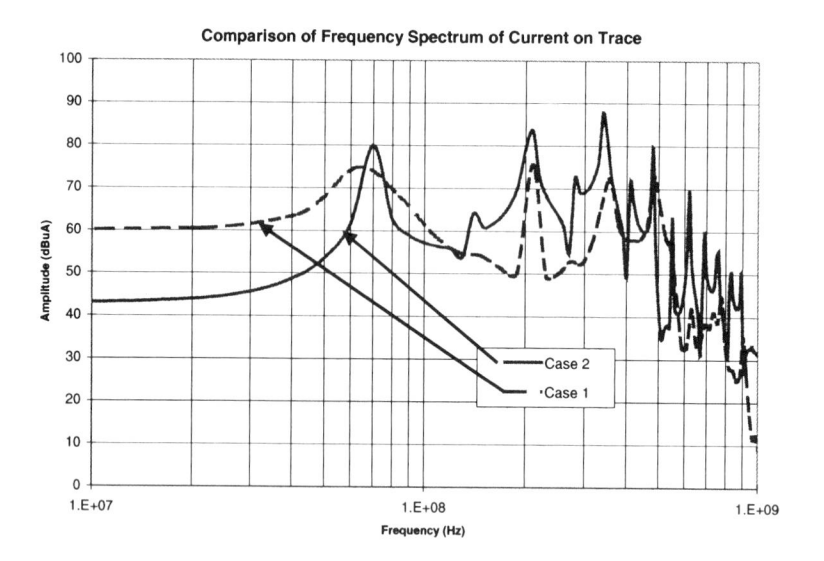

Figure 6-8 Current Waveform (Frequency Domain)

The emissions in the near field, just above the PCB will be a source of energy that can (and will) couple onto other internal wires (and ultimately be conducted to the outside of the metal enclosure).

These emissions will also be an energy source that can excite resonances within the metal enclosure and possibly leak out through slots, vent holes, and other apertures. Analysis of the near fields allows direct cause and effect analysis without the likelihood of masking the result due to some other effect, such as the resonant length of the external wires, etc.

6.5 Controlling Emissions from Intentional Signals – Loop-Mode

While emissions from this potential source are usually not the primary source of emissions, controlling emissions from this source is fairly straightforward. The primary strategy should always be to control the spectrum of the source signal. That is, do not create the high-frequency harmonic content of the current waveform unless it is truly required for proper circuit operation. Often the rise time of an intentional signal is much faster than required for functionality. A good design rule-of-thumb is that generally only the first five to seven harmonics of a signal are required to produce a good rise time. Higher frequency content will result in faster rise times, and likely higher costs for EMC emissions controls.

Once the signal harmonic content current amplitude is controlled, Equation 6.1 clearly shows that there is only one other way to control the emissions: to control the exposed length of the trace. High-speed traces should be routed on internal layers, buried between solid reference planes. Traces routed as strip lines (equal distances between the two reference planes), or as asymmetrical striplines are much preferred to exposed traces on the top or bottom layer of the board for high-speed signals. Naturally, some exposed portion of trace is required to connect to surface pads of components. These segments of the traces should be minimized and typically kept less than approximately a centimeter long.

6.6 Intentional Signals – Common-mode

Now we will discuss a different potential source than described in the previous sections. It is still directly caused by the intentional signal, but this time through an indirect radiation mechanism. The previous potential source assumed that all of the return current was contained directly under the microstrip trace. While most of the return current is under the microstrip trace, not all of it is confined to this narrow region. [6.4] The return current will spread outward on the reference plane to find the path of least impedance (dominated by inductance at high frequencies) back to the source. A simple microstrip trace is shown over a ground-reference plane in Figure 6-9. The current distribution across the ground-reference plane is shown in Figure 6-10.

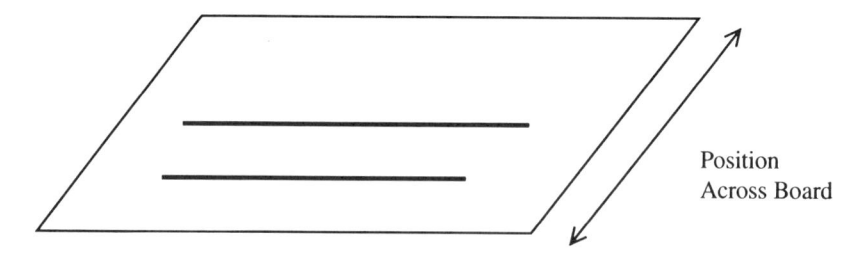

Figure 6-9 Two Microstrips Over Reference Plane

If the microstrip in the previous example was located relatively close to the edge of the ground-reference plane, the return current along the edge would be significant, and would cause radiation from that plane edge. When observing the board from the edge, the current on the ground-reference plane would resemble the current on a thin-wire antenna, similar to a dipole antenna. Again, the near fields are more useful to consider than the theoretical fields at 10 meters. The edge of the board is usually placed close to the metal enclosure, and will likely be near the seam of the enclosure, or an air vent area, etc. The near field emissions close to the edge (for example, within 2 inches) become the source that excites the seam slots, holes, and other apertures.

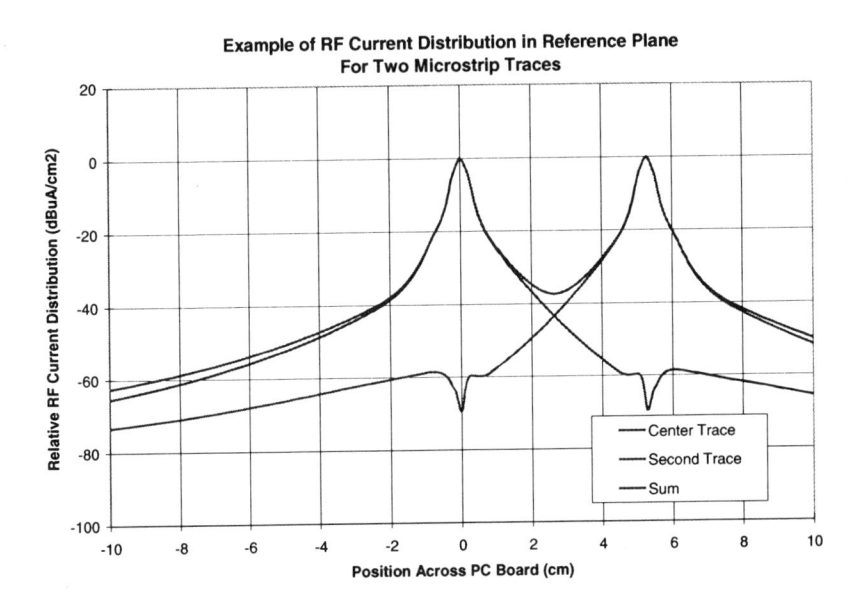

Figure 6-10 Current Distribution in Reference Plane

An example of the maximum electric field along the edge of a PC board due to the return current spread in the reference plane is shown in Figure 6-11. The microstrip is moved from very close to the edge to a few inches away. In this example, the board is 10 inches long, and the microstrip is 4 inches long. The edge of the board is scanned for the maximum electric field across a frequency range from 10-1000 MHz. As can be seen in this figure, the maximum electric field along the edge of the board is constant with frequency and increases as the microstrip is placed closer to the edge. Figure 6-12 shows a summary of the microstrip distance from the edge vs. amplitude of the electric field. The electric field amplitude changes very rapidly as the microstrip is moved from very close to about one-half inch from the edge. The field strength declines at a slower, linear rate for distances greater than one half-inch from the edge of the board.

98 / PCB Design For Real-World EMI Control

While it is impossible to prevent the return current from spreading out over the reference plane, the amount of emissions along the edge of the board can be reduced by keeping the microstrip traces as far from the edge as possible. A good rule of thumb is to keep all traces with high-frequency content that are parallel to the board edge at least one-half inch from the board's edge.

A second (and more effective) way to reduce the emissions from the edge of the board is to not create currents that contain harmonics in frequencies that are not required for functionality of the circuits. This is a technique that will work across all the various potential EMI sources.

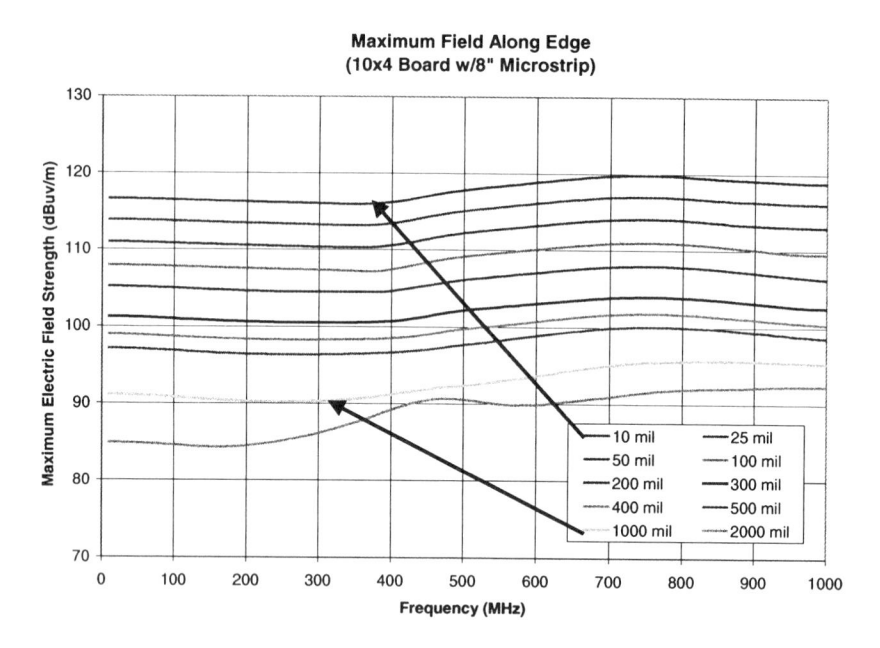

Figure 6-11 Maximum Electric Field Along Board Edge from Return Current Spread

6.7 Intentional Signals – Common-mode with Interrupted Return Path

As described in the previous section, the return current is contained *mostly* under the trace. This is true when the return current path is continuous. However, this return path is often not continuos. There are two major PCB design approaches that cause an interrupted return current path: critical signal traces crossing splits in the reference plane, and critical signals using vias between different layers within the PCB. Chapter 5 discussed the effects of interruptions in the return current path in detail. In this section, the effects will be highlighted in the context of this potential emissions source.

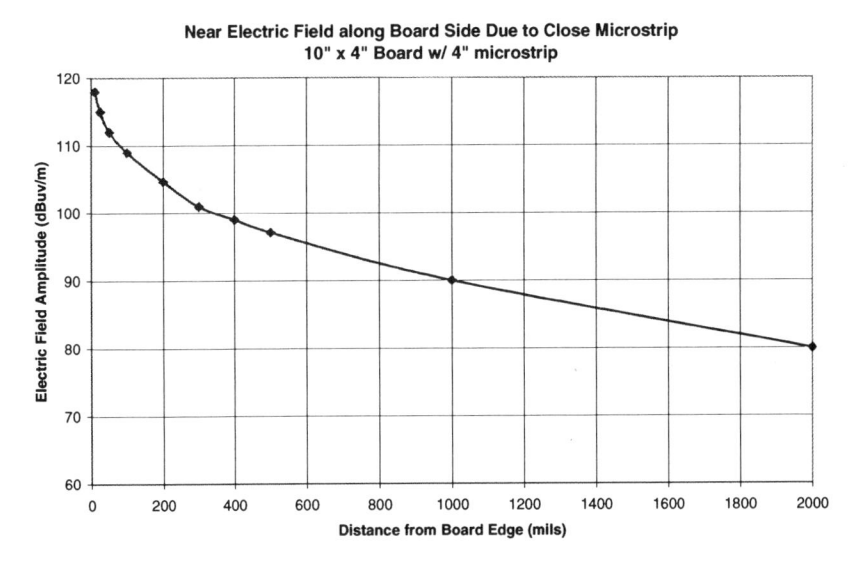

Figure 6-12 Electric Field vs. Distance From Edge

6.7.1 Critical Signal Traces Crossing Splits

Section 5-2 described the effects of traces crossing splits in their reference planes, as sometimes happens when the power plane is used as the reference plane. Different DC voltages are used on various plane islands to meet different IC requirements. If a high-speed trace is referenced to this DC power layer, it is likely to cross from one island to another, resulting in a split plane crossing.

Stitching capacitors are often used to provide a return current path across the split. They should be positioned immediately next to the trace that is crossing the split in the reference plane. Many times, however, there are numerous traces crossing in a small area (as with a high-speed data or address bus) and it is not practical to provide a stitching capacitor for each trace or each pair of traces. The farther away the stitching capacitor is placed (relative to the point where the trace crosses the split) the more common-mode voltage will be created across the split resulting in more emissions. [6.5]

Figure 6-13 shows how the common-mode voltage across the split increases as distance between the trace crossing point and the stitching capacitor increases. The common-mode voltage increases very rapidly for approximately the first one-half inch and then continues to increase at a slower rate. Placing the stitching capacitor within the first one-half inch will provide the greatest benefit.

Stitching capacitors can help lower the emissions from traces crossing a split in their reference plane, but only at low frequency harmonics. The natural inductance of the capacitor and the inductance associated with the connecting traces and vias greatly limit the effectiveness of stitching capacitors. For higher frequency harmonics, there is no effective way to reduce these emissions other than to not cross a split in the reference plane, and to use solid planes for high-frequency signals. Lowering the high-frequency harmonic content of the intentional currents will also reduce these emissions.

6.7.2 Critical Signals Through Vias

The previous section discussed a break in the return current path when a plane is used for more than one DC voltage. Interruptions in the return current path can occur vertically as well as horizontally. It is a frequent practice to route critical signal traces on multiple levels within the PCB. For complex boards, finding a path for all the signals (critical and non-critical) is difficult, and requires the use of different layers. With the proper design, these layer changes will not cause

EMC problems. Without the proper care, these layer changes can be one of the greatest contributors to EMC problems of all.

Increase in Common Mode Voltage Across Split Plane Due to Trace Crossing Split and Stitching Capacitor Location

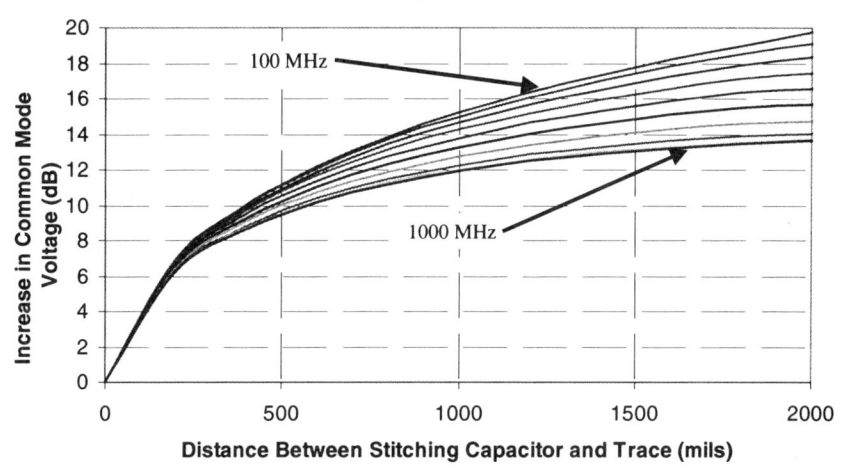

[Graph: X-axis "Distance Between Stitching Capacitor and Trace (mils)" from 0 to 2000; Y-axis "Increase in Common Mode Voltage (dB)" from 0 to 20. Curves labeled 100 MHz (upper) and 1000 MHz (lower).]

Figure 6-13 Common-mode Voltage Across Split Plane

Section 5.3 discusses the return current paths in detail when traces change their reference planes. When this occurs, the increase in return current path length causes increased electric fields between the two planes, which result in 'noise' voltage between the planes. This noise voltage will propagate across the PCB, cause emissions along the edges of the board, and potentially cause functional problems if the noise voltage is too great at sensitive IC's power pins. As discussed in Section 5.3, decoupling capacitors connected between the two planes can help with some of the lower to midrange frequency return currents. The return currents must use a combination of the decoupling capacitor path (conduction current) and a distributed capacitance path (displacement current) to travel from one plane to the other. The distributed capacitance is a path that is always present, but this path requires the current to travel over the entire board to use all the distributed capacitance. This results in long path lengths, which increases the path loop inductance, thus limiting the amount of effective distributed capacitance.

Using a via to change layers is not a bad practice by itself, changing the reference planes for the return current is the bad practice. Section 5.3 also discusses how PCB wiring layers can be changed using vias without causing return current path problems.

Some PC board stackup designs use many layers, and there may be a number of planes at the same DC voltage potential, such as "ground" or zero volts. This allows direct via stitching between the planes. A critical signal that is referenced to a "ground" plane on layer four (for example), and which changes its routing layer so that it becomes referenced to another "ground" plane (on layer 10, for example) could rely on the many vias between layers four and 10 for the return current path between layers. However, most of the return current will attempt to flow through the nearest via to the transition point. In order for other vias to have an effect, some of the return current must flow to those positions where the additional vias are located. If all the 'extra' vias are very close to the transition point (within a 0.5 inch radius), then some improvement in the return current path impedance can be expected. In general, using vias (or decoupling capacitors) farther away means that the return current must travel farther from the via transition point. This longer return current path causes additional common-mode currents in places on the PC board where they are not desired, and can also result in data pulse distortions.

6.8 Summary

There are a number of possible sources of EMC emissions. These potential sources are independent of each other, allowing us to consider how to best reduce the emissions from each of them in turn. This chapter discussed emissions from intentional signals. An intentional signal is a signal that is desired to be present at a given location for the functionality of the circuit. Chapter 7 will discuss emissions from unintentional signals.

One important concept in the discussion of all the types of sources is that nearly all the possible sources of emissions were due to the intentional signal, either directly or indirectly. If the intentional signals are properly controlled to contain only harmonics required for the functional signaling task, then the need to fight EMC emissions at higher frequencies is reduced or eliminated.

A second important concept is that most of the various potential emissions sources are directly related to the return current path, or the lack of a return current path. The most important EMC design consideration is to design this return current path explicitly. The design of the board layout should not simply be routing a trace from the driver to the receiver, but should include consideration of the return current path for the signal on that trace. This is probably the most important design consideration, since it affects so many possible emissions sources.

References

[6.1] F. Olyslager, E. Laermans, D. de Zutter, S. Criel. R. D. Smedt, N. Lietaert, and A. D. Clercq, "Numerical and experimental study of the shielding effectiveness of a metallic enclosure, *IEEE Trans. Electromagn. Compat.*, vol. 41, pp. 202-213, August 1999.

[6.2] M. Li, J. L. Drewniak, S. Radu, J. Nuebel, T. H. Hubing, R. E. DuBroff, and T. P. Van Doren, "An EMI estimate for shielding enclosure design," *IEEE Trans. Electromagn. Compat.*, vol. 43, pp. 295-304, August 2001.

[6.3] W. L. Stutzman and G. A. Thiele, *Antenna Theory and Design*, 2nd *Ed.*, John Wiley, New York, 1998.

[6.4] C.L. Holloway, G.A. Hufford, "Internal Inductance and Conductor Loss Associated with the Ground Plane of a Microstrip Line", IEEE Trans. Electromagn. Compat., V. 39, No.2, May 1997, pp. 73-77.

[6.5] R. Lyle, B. Archambeault, A. Roden, "Investigation of a Simplefied PCB System Using PEEC, MoM and FDTD," ACES Conference, Montrery Calif., March 2002

Chapter 7

Controlling EMI Sources – Unintentional Signals

7.1 Introduction

When starting to think about EMI design, the most effective approach is to consider the actual sources of EMI emissions, and treat them all individually. Most of the causes of EMI emissions at the printed circuit (PC) board level are separable and can be considered and treated individually, without increasing the emissions from some other source.

As discussed in Chapter 6, all signals on the board must be separated into two categories: intentional signals and unintentional signals. During the board design, engineers naturally consider the intentional signals. These are the signals they 'intend' to be on the board, and carefully design traces on the board to carry them from their source to their destination. The previous chapter discussed how to identify the intentional signal and the various sources of emissions from intentional signals.

Unintentional signals are often neglected during the design. After all, we don't 'intend' them to be on the board, so they are not often in our thoughts. Unfortunately, these unintentional signals are the cause of more than 90% of the EMI emissions from a PC board! Some amount of these unintentional signals will be present regardless of how carefully the board is designed. These unintentional signals must be considered and appropriate steps taken to insure they will not cause excessive emissions.

The sources of emissions can therefore be broken into two major categories: intentional signals and unintentional signals. Each of these major categories can be further broken down. As discussed in the previous chapter, sources of emissions from intentional signals include differential mode and common-mode sources. This chapter will discuss sources of emissions from unintentional signals including common-mode, crosstalk, power plane, and above board structures. Power/ground-reference plane noise (sometimes called simultaneous switching noise, or even [gasp] "ground bounce"[1]) is another form of unintentional signal source, but this subject will be discussed in a separate chapter.

7.2 Unintentional Signals

The previous chapter discussed various ways the intentional signal can cause EMC problems. The next few sections will discuss how *un*-intentional signals can also cause EMC problems. An unintentional signal is simply an intentional signal that exists someplace it was never intended to exist. All unintentional signals originate from an intentional signal which has somehow coupled onto a wire, trace, or conductor that was never intended to be a path for that signal.

Unintentional signals have the same frequency spectrum as the intentional signal, but since the signal is at a location never intended for it, these unintentional signals can be difficult to find and control unless they are understood and the proper design practices implemented to reduce their effects.

7.3 Unintentional Signals – Common-mode

In this section, the causes of the potential emissions from common-mode currents from *unintentional* signals will be discussed. Previously, in Sections 6.6 and 6.7, potential emissions from *intentional* signals with common-mode currents were discussed. In those sections, the return currents spread over the ground-reference

---

[1] "Ground bounce" is fairly common in the state of California, but is NOT a term that should ever be applied to electronic equipment unless the equipment is undergoing vibration testing…..

plane due to the natural inductance of the planes, or due to interruptions in the return current path. These same currents are the source of the unintentional signal common-mode currents.

For this potential EMI emissions source, the signal current on the ground-reference plane couples onto an I/O connector 'ground' pin and 'escapes' the shielded enclosure through the attached I/O cable[2]. Figure 7-1 shows an example where a critical signal trace is not directly connected to an I/O connector, however, the "ground" pin on the I/O connector is directly connected to the ground-reference plane. Return currents existing on the ground-reference plane are spread over the entire plane (as discussed in Chapter 6), and can easily couple onto the I/O connector's "ground" pin. Once this energy is on the I/O connector "ground" wire, it is connected directly to the outside of the shielded enclosure.

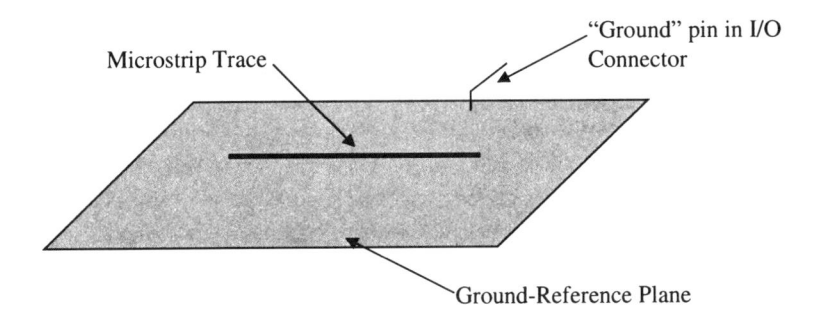

Figure 7-1 Example of I/O "Ground" Pin connected to Ground-Reference Plane

Most radiated emissions are caused by unwanted currents on external cables and wires. When a particular I/O cable is unshielded, any unwanted or unintentional currents on any of the conductors, even the "ground" conductor, will cause radiated emissions to increase. Even shielded cables are likely to have increased emissions, due to noise coupled onto the cable shield, unless care is taken to ensure a good (low impedance) electrical connection between the cable shield and the shielded enclosure chassis.

---

[2] Note: this section is NOT discussing signal I/O lines and how to filter them. Filtering I/O signal lines will be discussed in Chapter 9.

7.4 Controlling Emissions from Unintentional Signals – Common-mode

Section 6.6 described the spread of return currents in the reference plane, and the fact that this spread cannot be prevented where the reference plane is continuous. One way to keep the return current spread from continuing into the I/O connector area is to use an intentional split between the high-speed digital circuit area and the I/O connector area. Figure 7-2 shows an example of using such a split to isolate the I/O ground-reference area from the digital ground-reference area. This strategy can be very effective for isolating I/O "ground" pins from the return current spread.

A serious caution is needed here. A significant portion of the previous chapter was devoted to explaining why splits in reference planes are a "bad thing", and now we suggest that splits are a "good thing". The basic truth is that splits are neither "good" nor "bad" by themselves, the circumstances where they are needed must be clearly understood, and splits used only under the correct conditions. Allowing a high-speed trace to cross a split reference plane means that the return current must find some other return path, and this new return path will likely cause increased emissions. Reference plane splits should *never* be permitted adjacent to high-speed traces. When a low speed I/O area is near high-speed circuits, however, the low speed I/O connector "ground" pins can be effectively isolated from the high-speed return currents when using splits in the reference planes.

A good candidate for effective reference plane split use is a typical computer motherboard. There are a number of low speed I/O connectors on a typical personal computer motherboard, such as keyboard, mouse, serial port, parallel port, etc. All of these I/O ports have a 'ground' pin that is usually connected directly to the 'ground' plane on the motherboard. Any I/O port with an intentional data rate of less than 5 MHz can be considered a low speed I/O for these purposes. There is no need to have any high-speed traces (for example, clock traces, high-speed bus traces, etc) running near the low speed I/O connectors. Splitting the ground-reference plane in the low speed I/O area can be very effective at isolating the return currents that have spread out on the reference plane from the I/O connector 'ground' pins, and can be a low cost EMC design solution at the same time.

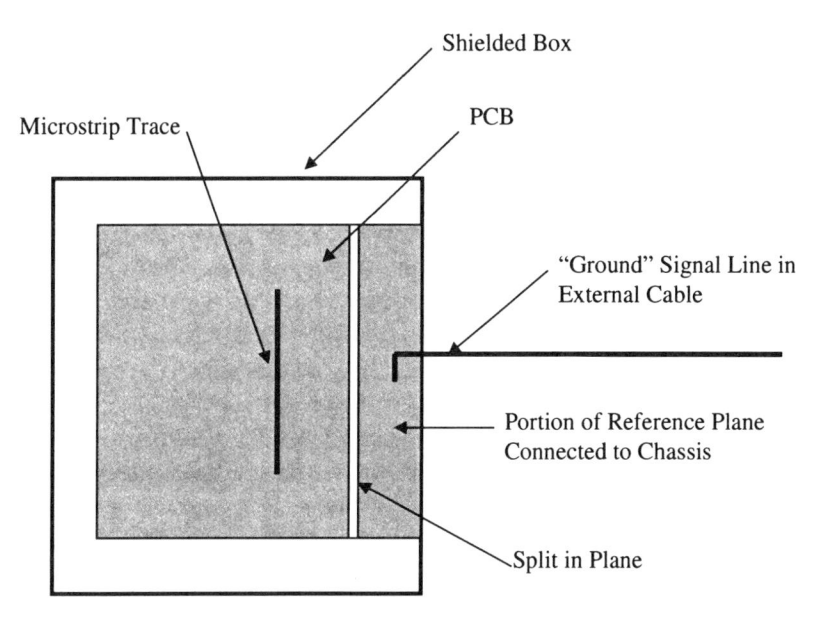

Figure 7-2 Example with Split Reference Plane

There are a couple of additional important design considerations to address in order for the split I/O ground-reference approach to be successful. First, the portion of the ground-reference plane that is split away from the main PC board digital ground-reference must be carefully connected to the shielded enclosure chassis with a low inductance (low impedance) contact. If this contact is omitted, is intermittent, or does not have a low enough impedance, the emissions can actually increase at some frequencies. Remember, the ultimate reference for any signals on the external wires and cables is the chassis shield. The I/O signal reference inside the enclosure near the I/O connector must be at the same potential as the external chassis.

The second important design consideration is the intentional low speed I/O data return current path. As discussed in previous

110 / PCB Design For Real-World EMI Control

chapters, current must always return to its source. The low speed I/O signals must have a return current path or data errors could occur, or the system could experience other failures. This return current path is provided by placing ferrite beads across the I/O split. The ferrite beads allow the return current to flow at low frequencies and block the high-frequency digital ground-reference return current spread. Capacitors should never be used across these splits, since a capacitor will allow the high-frequency noise to pass, and will block the intentional low frequency return currents. Figure 7-3 shows a diagram of the reference plane on a PC board with a split and with ferrite beads across the split. Since these beads are intended for low frequency return currents, only a few are needed across the board, distributed across the split near the low-frequency I/O connectors.

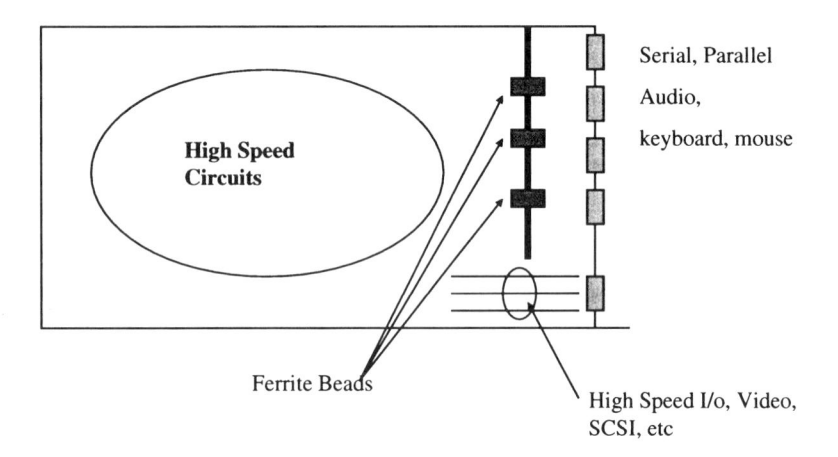

Figure 7-3 Split Reference Plane Example with Ferrite Beads Across Split

A split in the ground-reference plane is intended to provide a high impedance to high-frequency signals on the reference plane. The width of the split is not critical but should be 50 mils or more. Figure 7-4 shows the impedance across a split plane for an example PC board of 10" x 12". Note that at about 400 – 500 MHz the impedance becomes low due to a resonance effect. This means that using a split is effective in blocking the high-frequency current spread, but only up to a few hundred megahertz. At higher frequencies, the impedance of the split may not be as high as expected due to board and slot resonances.

Impedance (Magnitude) Across Split Plane

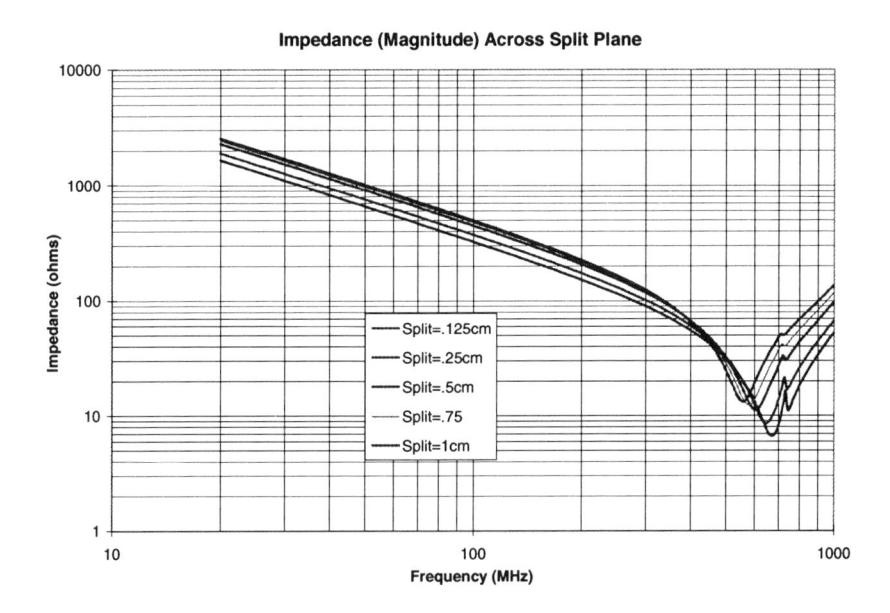

(Curves shown for Split=.125cm, Split=.25cm, Split=.5cm, Split=.75, Split=1cm)

Figure 7-4 Impedance Across Split in Reference Plane

An example of emissions from a simple metal enclosure with an internal PC board and a single external wire is shown in Figure 7-5. In this example, a microstrip trace is on the internal PC board, but does not come close to the I/O "ground" pin. The PC board's ground-reference plane is connected to the shielded chassis in the I/O connector area. The first plot shows the emissions from the enclosure with no wire connected to the I/O connector. Enclosure resonances are visible at about 425 MHz and 750 MHz. The external wire is added and resonances associated with the wire length are visible at 80 MHz, 225 MHz, etc. When a split is placed in the internal PC board's ground-reference plane between the microstrip trace and the I/O area, the emissions from the external wire are reduced by at least 10 dB for frequencies below 500 MHz. Above 500 MHz, it is difficult to see any improvement with the split. This

112 / PCB Design For Real-World EMI Control

is consistent with what is expected based on the previous impedance discussion.

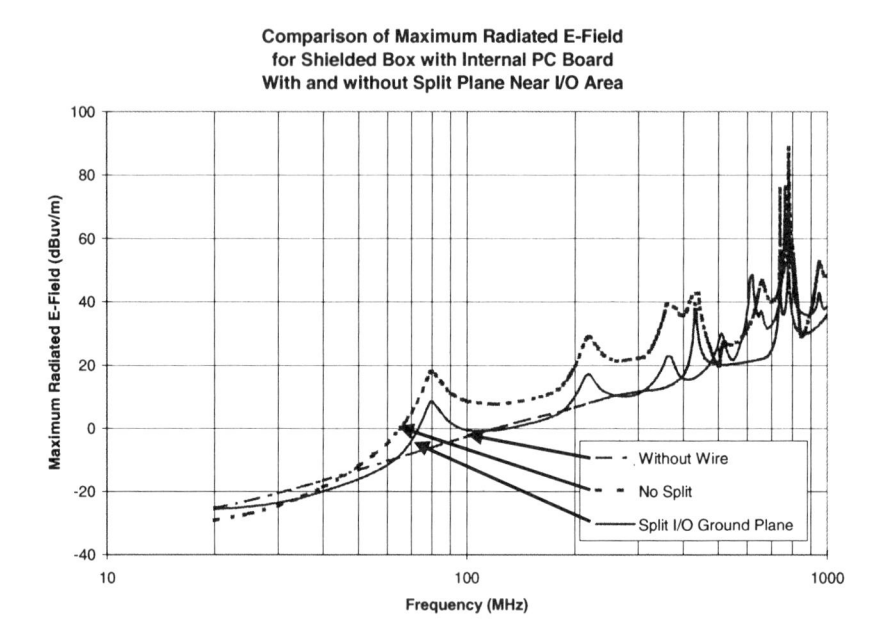

Comparison of Maximum Radiated E-Field for Shielded Box with Internal PC Board With and without Split Plane Near I/O Area

[Graph: Maximum Radiated E-Field (dBuv/m) vs Frequency (MHz), showing three curves: Without Wire, No Split, Split I/O Ground Plane]

Figure 7-5 Emissions From External Wire

For isolating frequencies above 400 – 500 MHz, the best technique is to eliminate the ground-reference plane near the low speed I/O connectors completely. The "ground" pins on the low speed I/O connectors should be treated as signal pins, and be filtered just as the signal pins are. In fact, the "ground" pins carry the same currents (if everything works correctly) as the signal pins. A ferrite bead in series between the 'ground' pin of the I/O connector and the digital ground-reference area provides a filter for the noise currents and provides a low frequency return current path. The ground reference plane is not present in the I/O area as shown in Figure 7-6.

As an added note, the power planes should also not be present in the I/O area since they are often used as reference planes for high-speed signals and will contain return currents which have spread out to find their own path of least impedance back to their sources. If these power planes are allowed into the I/O area, energy from these currents can couple onto the I/O signal lines (including the "ground" signal line) and be conducted outside the enclosure and cause emissions.

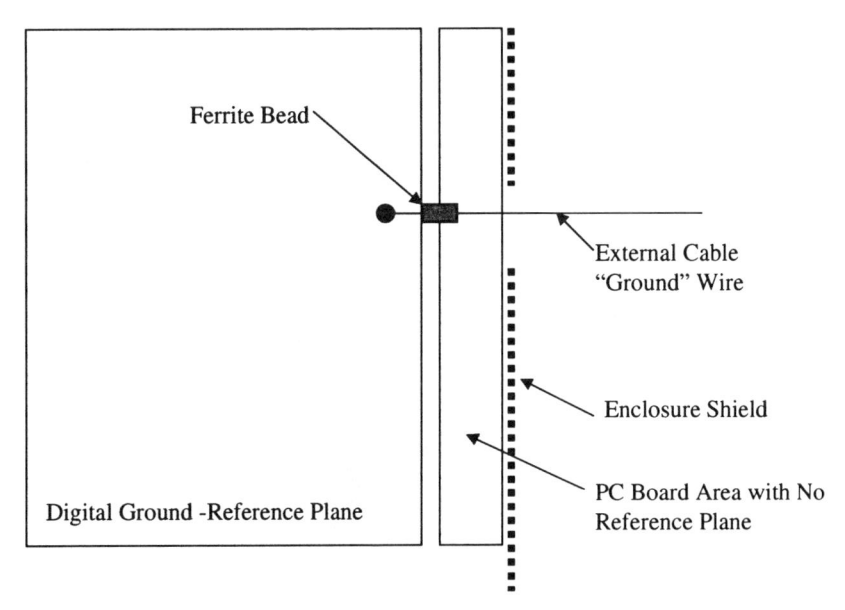

Figure 7-6 I/O "Ground" With Ferrite Filter

7.5 Unintentional Signals – 'Crosstalk' Coupling onto I/O Lines

'Crosstalk' coupling onto I/O lines is another means by which an intentional signal is coupled onto an unintended conductor, making it an unintentional signal on that conductor. Crosstalk is usually carefully monitored on critical signal traces during signal integrity analysis to insure that proper signal quality is maintained. From an EMC point of view, however, crosstalk between critical signal traces is not the concern, but crosstalk coupling between critical signal traces and I/O traces is an important focus.

Crosstalk coupling between a critical signal trace and an I/O trace is represented in Figure 7-7. The critical signal trace is not directly connected to the I/O trace, but they are routed close to each other, allowing crosstalk coupling to occur. The crosstalk coupling can be horizontal, as shown in the figure, or it could be vertical, i.e., coupling can occur between different internal layers within the PCB.

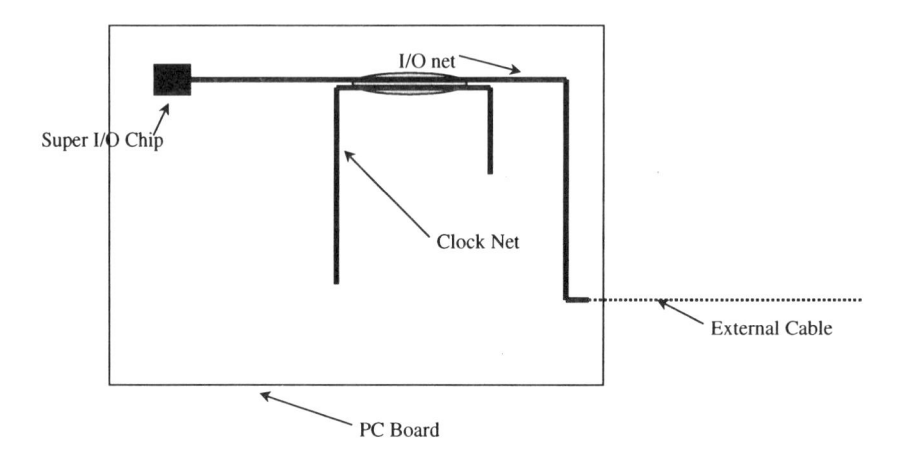

Figure 7-7 Single Level Crosstalk Coupling to I/O Trace

Multi-level, or cascaded crosstalk coupling is also a concern since very little current from the critical signal trace is needed on the outside of the shielded enclosure to cause unacceptable emissions. Figure 7-8 shows an example of multi-level crosstalk coupling. The critical signal trace is routed near an 'innocent' trace and some of the intentional signal currents are coupled onto the 'innocent' trace. The "innocent" trace is then routed close to an I/O trace, where secondary crosstalk coupling occurs, and a portion of the original intentional signal is coupled onto the I/O trace and becomes an unintentional signal.

The concern is to keep high-speed signal harmonics away from the I/O area, especially unshielded I/O connectors and cables. It only takes about 100 microvolts of common-mode noise on an external unshielded cable to cause emissions above commercial limits. [7.1]

Controlling EMI Sources – Unintentional Signals / 115

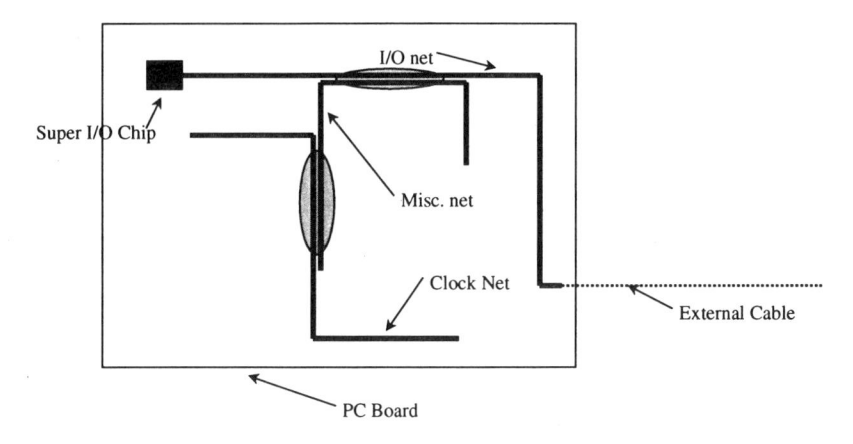

Figure 7-8 Cascade Crosstalk Coupling to I/O Trace

7.6 Controlling Emissions from Unintentional Signals – 'Crosstalk' Coupling to I/O Lines

Obviously, the best way to control crosstalk coupling to I/O lines is to keep high-speed traces and I/O traces far away from each other. The most effective way to isolate these types of traces is to route the two types of signals on different layers in the PC board stackup with a solid plane between them. It is often overlooked that crosstalk can and does occur between different layers when there is no solid plane between them. Figure 7-9 shows an example where the low speed I/O signal traces are routed on the outside layers of the PC board, and the high-speed signals are routed on the inner layers. Note that the high-speed traces are not exposed on outer layers (as discussed in Chapter 6) and the high-speed trace does not change reference planes (again, from Chapter 6). This strategy is effective for other concerns as well as the crosstalk coupling to I/O traces concern.

Unfortunately, it is not always possible to keep I/O signals and high-speed signals separated by solid planes. The desire for some separation between traces is still valid but often hard to achieve because of the need to utilize all routing channels on the PC board. Figure 7-10 shows a diagram of a PC board with a driven (high-speed) trace, a susceptible (I/O) trace and adjacent wiring channels.

116 / PCB Design For Real-World EMI Control

In this arrangement, it is recommended that the I/O trace be moved away from the high-speed trace, leaving empty wiring channels between them.

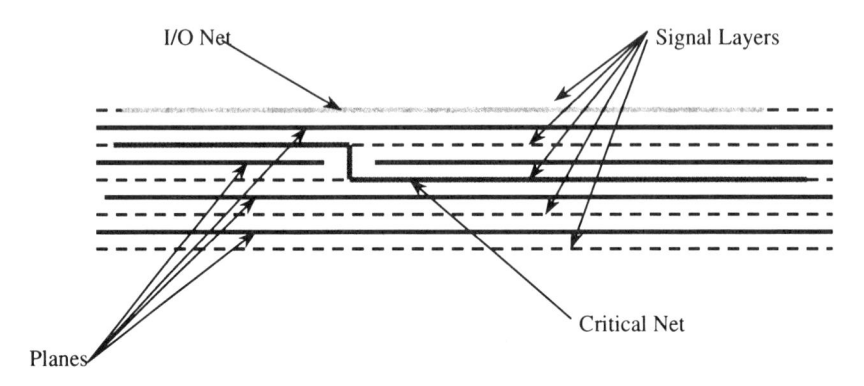

Figure 7-9 Isolation Using Solid Planes

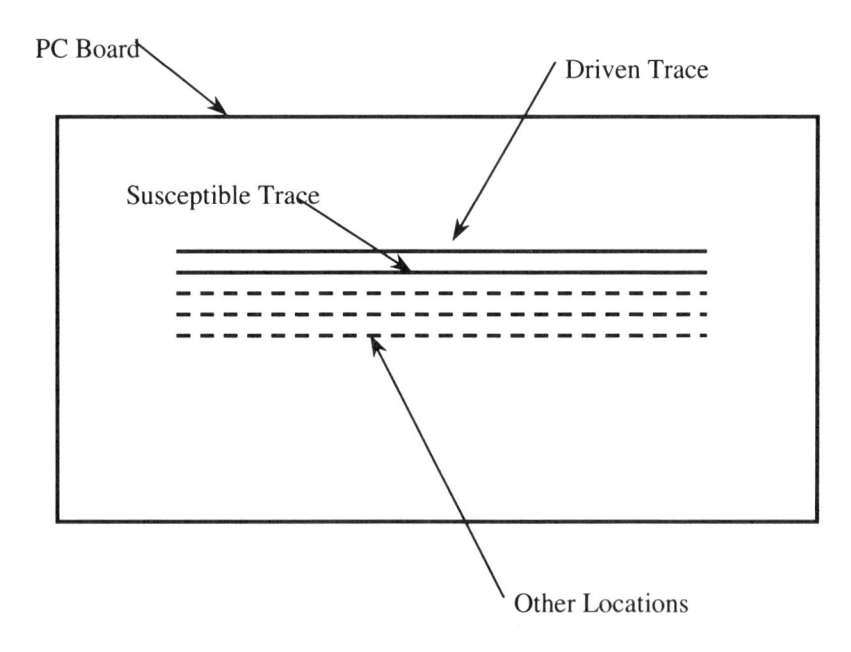

Figure 7-10 Crosstalk Coupling Example

A more effective way to isolate traces on the same layer is to use a guard trace. A guard trace is connected to the ground-reference plane by vias every inch (or more often)[3] and positioned between the high-speed trace and the susceptible trace. If there are a number of associated high-speed traces, such as in a bus, then they can be routed together with guard traces routed on the perimeter of the entire bus. Figure 7-11 shows the same PC board configuration as shown in Figure 7-10, except that a guard trace has been added.

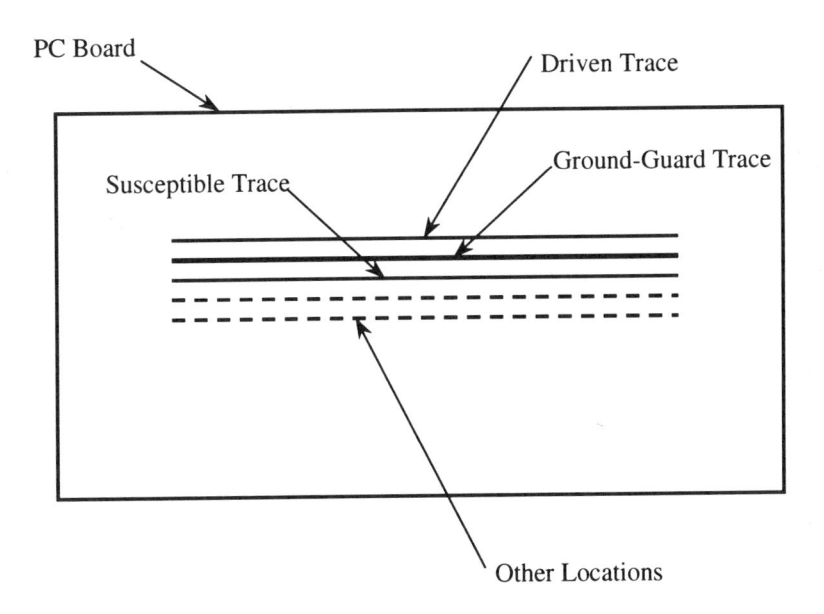

Figure 7-11 Trace Positions With Ground-Guard Trace for Isolation

Figure 7-12 shows the additional loss that physical separation or a guard trace provides over having two adjacent traces. In this plot, 0 dB indicates the same amount of crosstalk coupling as the case with two adjacent traces. As can be seen in this figure, the physical

---

[3] The spacing between the vias for a ground-guard trace should be less than 1/10th of a wavelength at the highest harmonic frequency contained in the intentional signal. For 1 GHz, a one-inch spacing is effective.

separation provides a few dB of additional isolation, but the guard trace provides the best isolation, and uses less board real estate.

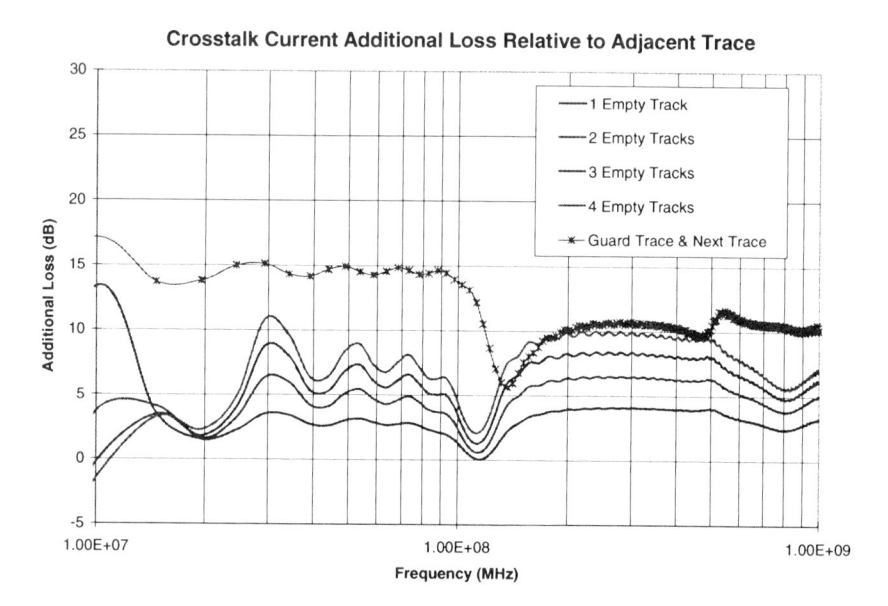

Figure 7-12 Loss from Physical and Ground-Guard Trace Isolation

7.7 Summary

There are a number of possible sources of EMC emissions. These sources are independent of each other, allowing us to consider how to best reduce the emissions from each of them in turn. This chapter discussed emissions from unintentional signals. Chapter 6 discussed emissions from intentional signals. Both intentional and unintentional potential sources must be considered during the design.

An important theme throughout the discussion of unintentional sources is that all the possible sources of emissions were due to the intentional signal. If the intentional signals are properly controlled to contain only harmonics required for the functional signaling task, then the need to fight EMC emissions at higher frequencies is reduced or eliminated.

References

[7.1] C. R. Paul, *Introduction to Electromagnetic Compatibility*, Wiley Inter-Science, New York, 1992.

Chapter 8

Decoupling Power/Ground Planes

8.1 Introduction

It is likely that decoupling power and ground planes is one of the most misunderstood design concepts, and certainly the area having the most myths about the 'correct' decoupling strategy. The use of decoupling capacitors connected between the power and ground planes on a printed circuit board (PCB) is a common practice to help ensure proper functionality(i.e. signal integrity) and to reduce EMI emissions from printed circuit boards. The proper number and value of decoupling capacitors is always a topic of debate between EMC engineers and design engineers. Some typical rules-of-thumb include requiring a decoupling capacitor for each power pin on an IC, at least one decoupling capacitor per side of physically large ICs, and/or decoupling capacitors spread evenly over every square inch of the board. Few qualitatively proven approaches for the optimal approach to decoupling is available in the technical literature. These rules-of-thumb can often result in drastic over-design of the decoupling strategy, since the saying 'better safe than sorry' is often applied. Many of these rules are really based in myth. In addition, there are a number of outright myths that exist and are published, causing significant confusion within the general design community. Some of these myths have some level of rationale justifying them; others do not. One published article claimed that the decoupling capacitors actually *caused* the emissions! Unfortunately, many myths seem infinitely plausible and are not easy to discount.

Traditionally, the values of the decoupling capacitors are largely based upon habit and the experience of the EMC engineer. Values of 0.01 uF or 0.1 uF are typically used. Often smaller capacitors are used in parallel with the main decoupling capacitor to provide a high-frequency and a low-frequency filtering effect. By using multiple capacitor values in close proximity, however, there is a risk of causing cross resonances that can have an adverse effect on noise and emission levels.

The overall result is that a design approach for the power plane decoupling (between a power plane and a ground reference plane) has historically been difficult to develop or analyze. With on-board clock speeds of 400 – 800 MHz becoming common, a more rational approach must be taken to optimize the design of decoupling capacitors on the printed circuit (PC) board.

8.2 Background

There are two primary purposes for using decoupling between power and ground-reference planes. The first purpose is for functionality, that is, the decoupling capacitor is a charge storage device, and when the IC switches state and requires additional current, the local decoupling capacitor supplies this current through a low inductance path. If the capacitor is able to supply all of the current required by the IC, then the voltage at the IC power pin remains constant at the desired supply voltage. If the capacitor is not able to supply the required current, then the voltage at the IC power pin is lowered temporarily until adequate current is provided, or until the need for the current is ended. If sufficient current is not provided, the IC may experience a functional failure. It is important, therefore, to locate decoupling capacitors close to the demand for current (IC power pins). It is also important to provide a low impedance path from the IC power pins to the power plane, from the IC's ground-reference pins to the ground-reference plane, and from the decoupling capacitor to the power and ground-reference planes. Capacitors used for this charge delivery function must have low equivalent series resistance (ESR) and low equivalent series inductance (ESL). Table 8-1 shows typical values for different SMT capacitor types.

Type	Package	ESL	ESR @ 100 MHz
NPO	0603	0.6 nH	60 mΩ
	0805	1 nH	70 mΩ
	1206	1 nH	90 mΩ
X7R	0603	0.6 nH	90 mΩ
	0805	0.9 nH	110 mΩ
	1206	1.2 nH	120 mΩ
Y5V	0603	2.5 nH	80 mΩ
	0805	3.1 nH	90 mΩ
	1206	3.2 nH	100 mΩ
X5R	0603	0.4 nH	60 mΩ
	0805	1 nH	80 mΩ
	1206	1.1 nH	110 mΩ

Table 8-1 Typical Values for SMT Capacitors

The second purpose for decoupling capacitors is to reduce the noise injected into the power and ground-reference plane pairs and thus reduce the EMI emissions from the edge of the circuit board. For example, the edge of a board may be near the seams of the metal enclosure or near an air vent area, allowing this noise to escape the enclosure. Another possibility is for this noise to couple onto I/O connector pins and be directly coupled out of the metal enclosure through any of the cables. There are a variety of coupling mechanisms that are possible once this noise is created.

The source of this injected noise can be either: (1) the temporary lowering of voltage at the IC pin power pin when sufficient current is not provided from the decoupling capacitor (creating a short duration voltage pulse), or (2) the noise signal injected between the power and ground-reference planes due to an intentional current (e.g. a clock signal or other fast switching signal) transitioning to different PC board layers on a via. Measurements on active ICs have shown that the relative magnitude of typical power and ground-reference plane noise sources is about the same. [8.1]

Once a signal from either of these sources results in noise between the power and ground-reference planes, the goal is to reduce the noise voltage level using decoupling capacitors.

8.3 Calculating the Source of Decoupling Noise

Since there are two different sources of this decoupling noise, the magnitude of the noise must be calculated differently for each source. While the cause of the noise is different, studies have shown that the relative level of noise is similar for both potential sources. Both sources must therefore be considered. [8.2]

8.3.1 Decoupling Noise from ASIC/ICs power pins

While it is common knowledge that some ASIC/ICs cause more noise problems than others, it has been traditionally difficult to predict the amount of noise from the ASIC/IC. First, consider how this noise is created. The ASIC/IC is not a generator where some noise exists internally, is conducted out of the ASIC/IC through the power/ground pins, and then couples into the power/ground plane pair. The ASIC/IC can be considered a fast switch. The impedance of this switch changes from a high impedance to a low impedance very quickly. In a low impedance state, the ASIC/IC is trying to draw more current from the supply. If the supply is able to meet this current demand, there is no problem, and no noise. In some cases, however, the supply cannot provide the current as quickly as the ASIC/IC requires it (mostly due to inductance) and there is a momentary decrease in the supply voltage locally at the ASIC/IC pin.

This lowered voltage results in a noise pulse that will propagate around the power/ground plane pair, and is commonly known as "decoupling noise".

It is necessary to focus on the current the ASIC/IC is trying to draw, and not a noise voltage, when trying to predict the amount of noise to expect between a power/ground plane pair since this current is the actual cause of the noise. The resulting noise voltage is very dependent on the impedance of the power/ground plane pair at the point where the ASIC/IC power pins are located, and across the entire frequency range of interest.

8.3.1.1 ASIC/IC Power Current Requirements

There are two main types of current required by an ASIC/IC. First, the ASIC/IC is likely to have some I/O drivers. These I/O drivers have significant current drive requirements and must draw this current from the power supply. The second required current is often called "shoot through" current, and is not used in the I/O drivers of the ASIC/IC, but rather is used within the ASIC/IC for internal switching and processing.

Depending on the type of ASIC/IC, either of these types of current is likely to dominate the power requirements. An ASIC/IC with a lot of I/O, such as a clock buffer with a large number of outputs or a bus controller, will have the I/O current requirements dominate the power current demand. An ASIC/IC with only a few I/O lines will likely have the shoot-through current dominate. Both types of current can be considered individually, and simply added together to find a better estimate of the total current.

8.3.1.2 Finding the ASIC/IC Power Current Demand for I/O

The I/O current can be found conveniently by using standard signal integrity tools. Normally, signal integrity tools are used to simulate voltage waveforms on various transmission lines. Many of these tools can also be used to find the current on the I/O trace. For clock buffers, the total I/O demand current is simply found by multiplying the individual trace current by the total number of active I/O lines. This is the most accurate technique to determine the pulse shape and allow accurate determination of the frequency domain harmonic content of the current pulse.

If it is difficult to obtain the current from a signal integrity tool, another simple technique is to use a simple capacitive load method given in

$$I_L = \frac{C_L n V_{CC}}{t_r} \qquad (8.1)$$

where:
C_L = the Capacitive load, normally about 10 pF for CMOS devices,
n = the number of active loads,
V_{CC} = the supply voltage, and
t_r = the rise time of the output pulse.

The current pulse shape can be approximated as a simple triangular pulse, and the peak amplitude of this pulse is found from the above method. The width of the pulse is assumed to be equal to the rise time plus the fall time of the I/O signal. While not as accurate as the signal integrity tool method, this method provides reasonable estimate of the current demand.

8.3.1.3 Finding the ASIC/IC Power Current Demand for "Shoot Through"

For clock buffers and other ICs, the peak amplitude of the shoot through current can be estimated using a vendor data sheet parameter called power dissipation capacitance (C_{pd}). The same triangle waveform shape is assumed, and the current pulse amplitude is found using

$$I_{p2} = \frac{C_{pd} * m * V_{cc}}{\Delta t_2} \qquad (8.2)$$

where:

m = number of I/O drivers, and
$\Delta t_2 = t_r + t_f$, and
t_r and t_f are the rise and fall times, respectively.

The total IC power current is the sum of the shoot through current and the I/O current. The waveform for the total power current is shown in Figure 8-1.

This method of using the I/O load current and the C_{pd} parameter to predict the noise voltage at the IC power pins was applied to a few standard clock buffers. These same ICs were used on a simple PC board and their power bus noise was measured. Figures 8-2 and 8-3 show the results for two different clock buffers. The agreement between the predicted and measured is quite good for such a simple calculation.

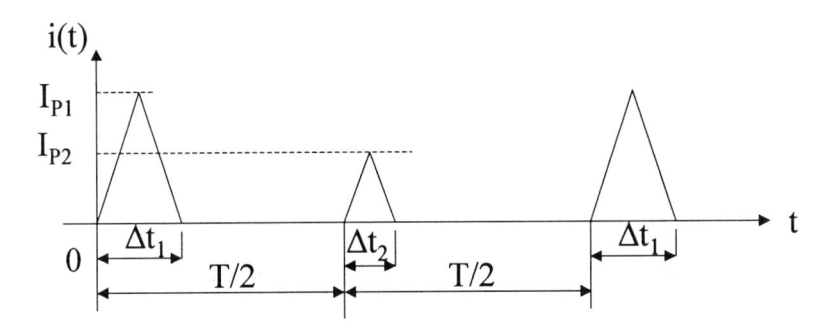

Figure 8-1 Power Pin Current Waveform for ASIC/ICs

8.3.1.4 Finding the ASIC/IC Power Current Demand without C_{pd}

The C_{pd} parameter is available for many clock buffers and other ICs, but not all ASIC/IC manufacturers specify or document this parameter. Depending on the complexity of the ASIC/IC, a first order approximation can be made using the I/O drive current only. The assumption that the I/O current dominates depends on the number of output drivers. On devices with a high number of outputs, this assumption can give a reasonably good prediction. Figure 8-4 shows an example of the prediction and measurement for a clock

128 / PCB Design For Real-World EMI Control

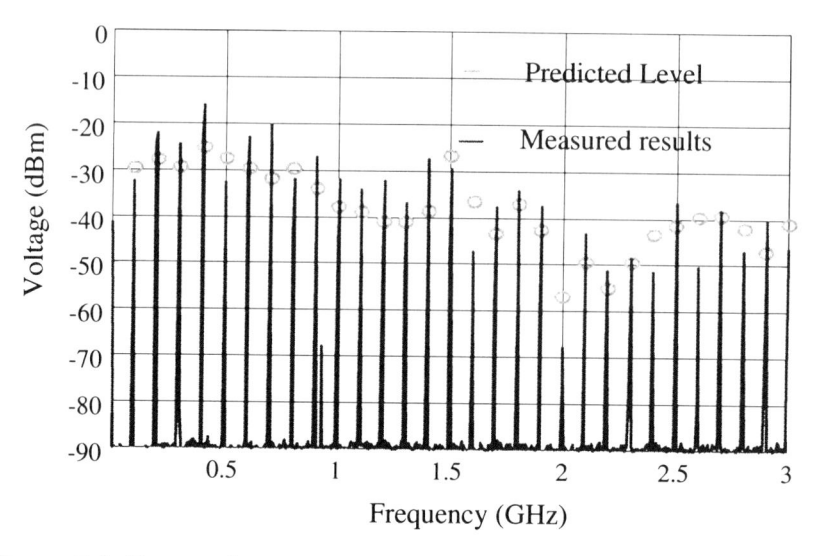

Figure 8-2 Expected and Measured Levels for the MPC905 Clock Buffer

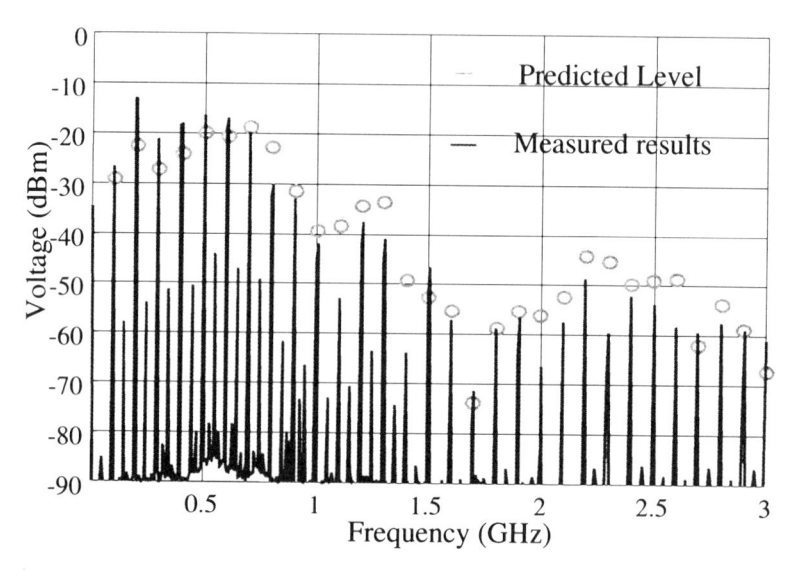

Figure 8-3 Expected and Measured Levels for the MPC946 Clock Buffer

buffer where no C_{pd} value was available from the manufacturer. The agreement between measured and predicted values is not as accurate as in the previous examples, but it is still an adequate estimation.

Large ASICs, such as processors, memory controllers, bus controllers, etc., require the manufacturer to do extensive simulations just to insure the device operates properly. Major ASIC manufacturers, such as IBM, Intel, ServerWorks, AMD, etc., create SPICE models to simulate the ASIC under various load and operational conditions. Most of these SPICE models are developed for signal/data flow in the ASIC, however, SPICE models are also created for the power requirements of the device. While ASIC vendors are seldom asked for these models (or the model results) from companies using these devices, these models can provide a very good indication of the power current requirements for large ASIC devices.

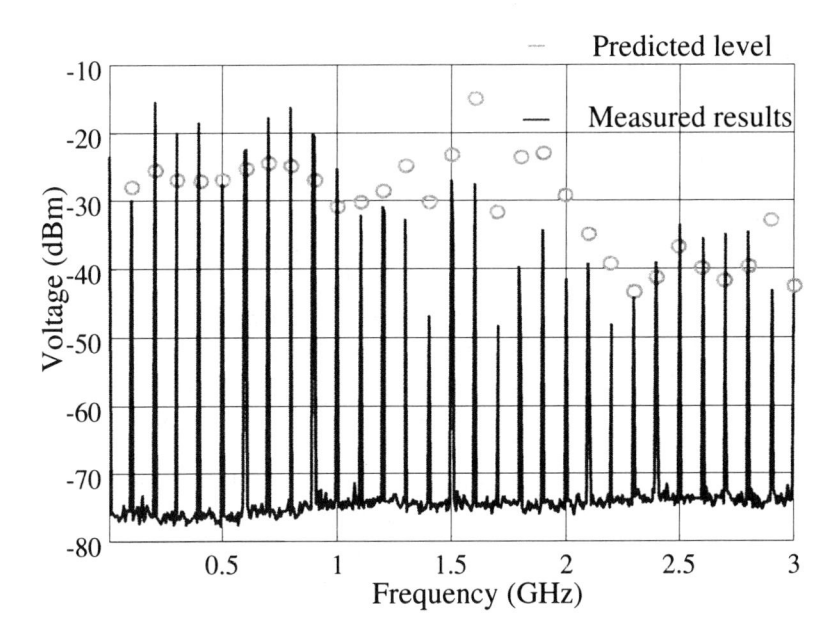

Figure 8-4 Expected and Measured Levels for the IDT807 Clock Buffer Using I/O Current Only

8.4 Decoupling Capacitor Effectiveness

The previous section discussed ways to determine the level of the source of noise between the power and ground-reference planes from an ASIC/IC. Regardless of this level, design practices often attempt to reduce the amount of this noise voltage some distance from the source of the noise by using decoupling capacitors. A good measure of decoupling "goodness" is to use a transfer function of the voltage at the observation point relative to the source voltage. This transfer function is dependent on both frequency and observation point. The definition of the transfer function (transfer impedance) is shown in (8.3). The transfer function relates the current at the source (the IC power pin) to the voltage at the observation point (some distance away). Since the goal is to have the least amount of noise at the observation point, we want Z_{21} to be as low as possible.

$$Z_{21}(f) = \frac{V_o(f)}{I_s(f)}\bigg|_{I_o=0} \quad (8.3)$$

While (8.3) is a simple expression, calculating this transfer function across the frequency range of interest, including the important parameters such as the inherent capacitance of the parallel planes, the inductance of the vias connecting the capacitors, the internal parasitic parameters of the capacitors, and the physical separation between capacitors is an extremely difficult task. The transfer function will vary depending on the size of the PC board and when any of the previous values are changed, but we can make some general predictions using a standard size PC board. A test PC board was created to help illustrate the effects of a variety of decoupling capacitor configurations. Measurements of the transfer function were made on this test board to show the effectiveness of each capacitor configuration.

8.4.1 Test Board Description

This example focuses on a 4-layer PC board with external dimensions of 10" x 12", and is considered a typical board size[1]. The PC board stackup consists of a top layer with connection/solder pads, two solid planes (a "power" and a "ground" plane), and an unused bottom layer. The separation between the solid planes was 0.035". Since the frequency range under investigation extends from 30 MHz to over 1 GHz, a repeatable and well-controlled connection method to the test board is required. A series of 15 SMA connectors were mounted across the board as shown in Figure 8-5. The center pin from each SMA connector was connected to the lower plane, and the outer conductor of the SMA connector was connected to the upper plane. Each of the SMA connectors are surrounded by four locations for SMT (0805 size) decoupling capacitors. Figure 8-6 shows the detail of each location. The PC board dielectric was standard FR-4 material with a relative dielectric constant of 4.5.

In addition to the SMA connectors and their local decoupling capacitor sites, locations for additional SMT decoupling capacitors are located on a one-inch grid on the PC board, as shown in Figure 8-7.

An equivalent circuit of the PC board is shown in Figure 8-8. The 99 capacitors are each represented with an equivalent circuit including its capacitance (C_x), equivalent series resistance, and equivalent series inductance. The inherent capacitance of the two parallel planes (C_{planes}) is in parallel to all the discrete capacitors. Note that this is only a low-frequency model and does not include resonance effects due to the physical size of the board (which will dominate at high frequencies).

---

[1] Note: the board size will affect the resonance frequencies.

132 / PCB Design For Real-World EMI Control

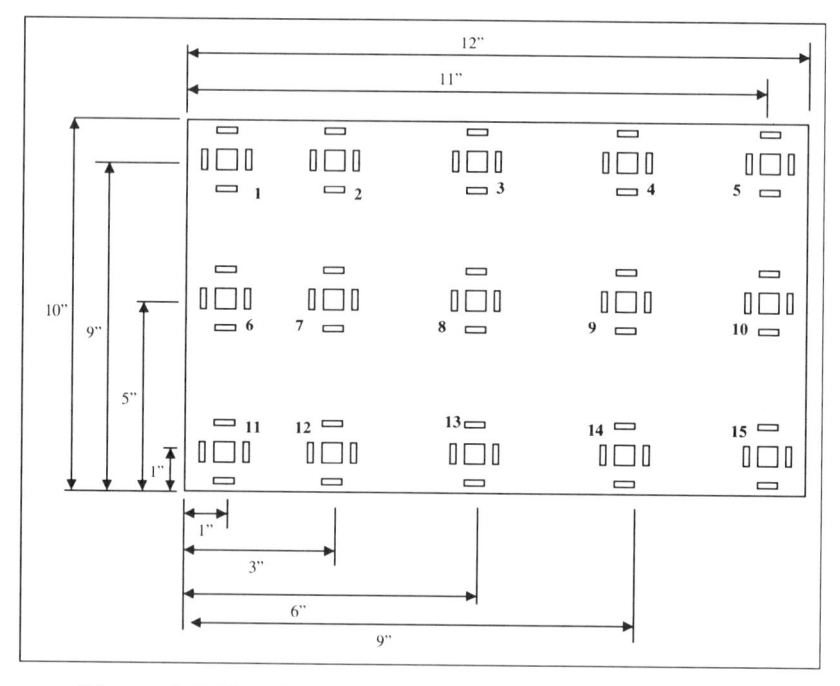

Figure 8.5 Test Board SMA Connector Configuration

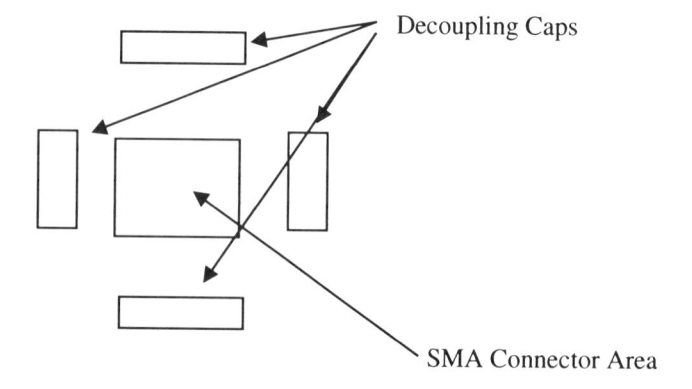

Figure 8-6 Test Board SMA Connector Area Detail

Figure 8-7 Additional Decoupling Capacitor Locations on Test Board

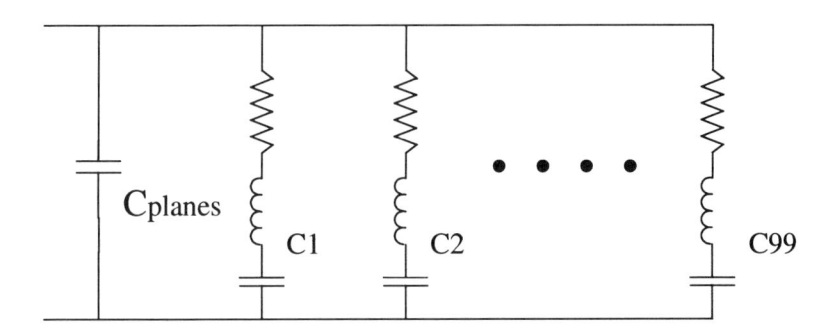

Figure 8-8 Equivalent Circuit for Decoupling Test Board

The physical size-based resonances will begin at frequencies where the cavity created between the two metal plates is one-half wavelength (or integer multiples of half wavelengths). The electric field is assumed to not vary in the z-direction (normal to the plates). This size-based resonant frequency can be found from

$$f_{mn} = \frac{1}{2\sqrt{\varepsilon\mu}}\sqrt{\left(\frac{m}{a}\right)^2 + \left(\frac{n}{b}\right)^2} \tag{8.4}$$

where:
m and n are the mode number (only 1 can be zero)
and a and b are the dimensions of the PC board.

All measurements were taken with a network analyzer. The transfer function was found using S_{21} two port measurements where

$$|S_{21}| \cong \frac{|Z_{21}|}{Z_0} \tag{8.5}$$

where Z_0 is the 50 Ω characteristic impedance of the test instrument.

8.4.2 Empty Test Board Configuration

There were a number of different decoupling configurations evaluated, as well as different capacitor values. The difference between local source decoupling and distributed decoupling was also investigated. Finally, the number of capacitors and various values of capacitance (both single and multiple values) were examined.

Throughout the measurements of the S_{21} transfer function, significant resonant effects are found at high frequencies. Figure 8-9 shows the transfer function from the center of the example test board to one corner. While the exact values of the transfer function will change as the source and observation locations change on the PC board, the general effects will be consistent. As apparent in Figure 8-9, resonances due to the board dimensions become the dominant factor above 200 MHz. Depending upon which location on the PC board was used for the transfer function measurement, a particular resonant frequency may or may not be excited. From an EMC point of view, however, it is not possible to control where a component might be placed. To ensure a particular resonance mode would not be excited, all modes must be assumed present (or a worst case "envelope"). For this set of experiments, the center and one corner port were used so that most of the resonant modes would be excited over the frequency range of interest (< 2 GHz).

Further experiments with different capacitor values at various locations demonstrate that while a particular resonance would shift to a different frequency (as capacitors were added and/or moved about the board), the general shape and impedance of the resonant peaks did not change. This indicates that the entire frequency range must be considered resonant. That is, since the frequency of resonance shift with any change in placement of value of the capacitors, all frequencies must be assumed to be resonant, and the overall envelope of the S21 transfer function lowered. These resonances are extremely important and require a full-wave electromagnetic analysis to consider the physical dimensions of the board. This resonance variation makes SPICE circuit analysis ineffective for high-frequency decoupling analysis.

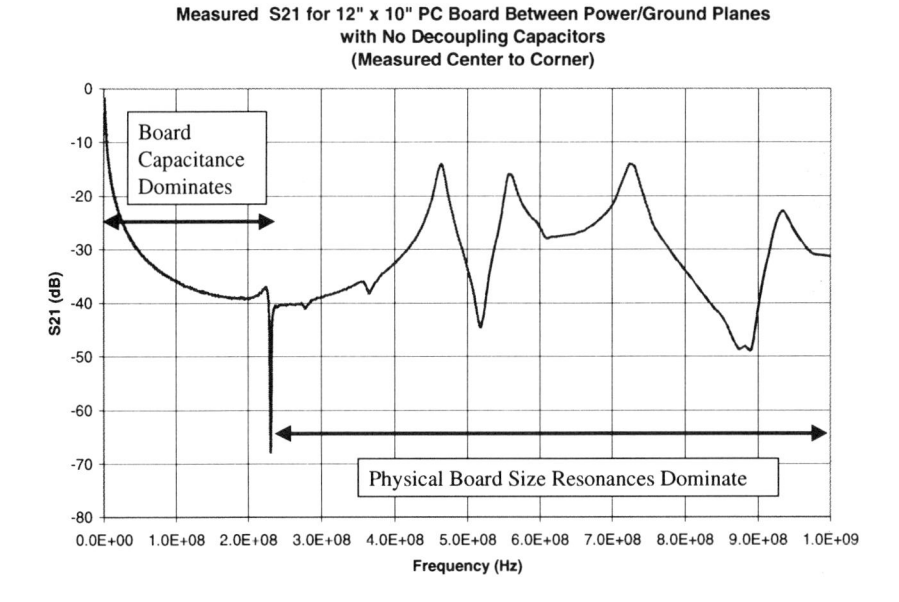

Figure 8-9 Transfer Function for Board with No Decoupling Capacitors

8.4.3 Quantity of Distributed (Global) Decoupling Capacitors (.01uf Only)

When decoupling capacitors are distributed evenly across the entire PC board they are considered *global* decoupling capacitors. They are not intended to provide decoupling to any specific IC, but instead to the entire PC board. The number of decoupling capacitors and the density of their placement is important to maintain a low transfer function across the board.

For the following example, all the capacitors used are 0.01 uF[2] SMT[3]. To illustrate the effect of different capacitor densities, the example 10" x 12" board was populated with 25, 50, or 99 (every inch) capacitors. In each case the capacitors were evenly distributed across the board. The results and conclusions for the various board source and receive locations were consistent at high frequencies. While the individual resonances might shift slightly, the overall envelope of the S_{21} transfer function is not affected by location.

As can be seen in Figure 8-10, adding capacitors lowers the S_{21} transfer function in the lower frequency ranges (below 200 – 400 MHz). At higher frequencies, the overall envelope for the S_{21} transfer function decreased only very slightly (despite a resonant frequency shift) as more capacitors were added.

The maximum number of capacitors populated was 99 distributed capacitors, representing one capacitor every square inch across the PC test board. This is a much more dense capacitor placement than is often possible on real products. Even with this dense capacitor distribution, the decoupling transfer function was only improved below approximately 400 – 600 MHz. This frequency limitation is due to the connection inductance of the vias, traces, pads, etc., which are inherent in the capacitor attachment.

---

[2] A variety of capacitor values were selected for this study.
[3] Surface mount style capacitors in the 0805 package size were used throughout this study.

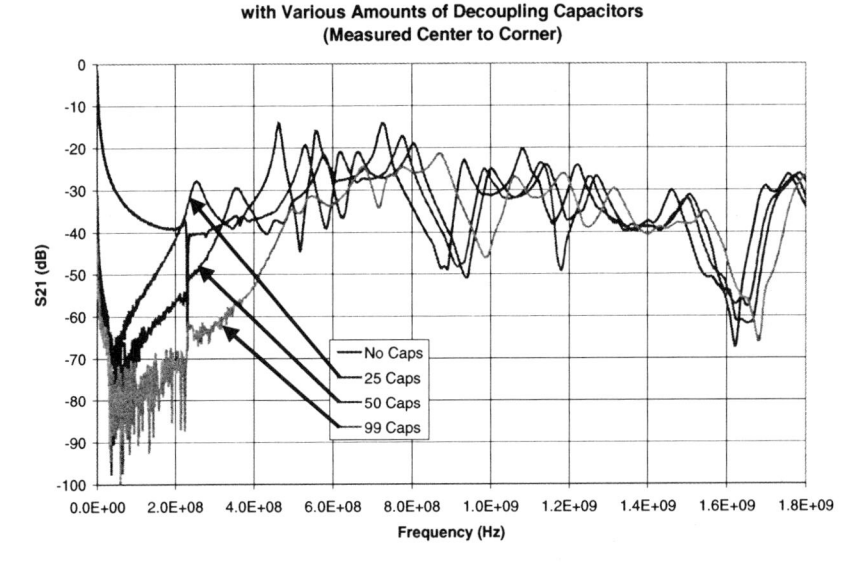

Measured S21 for 12" x 10" PC Board Between Power/Ground Planes with Various Amounts of Decoupling Capacitors (Measured Center to Corner)

Figure 8-10 Transfer Function with 0.01 uF Decoupling Capacitors

8.4.4 Quantity of Distributed Decoupling Capacitors (0.01uf and 330 pF)

A seemingly obvious solution to the limited high-frequency performance is to add capacitors with lower impedance at high frequencies. Using a combination of so-called "high-frequency" capacitors and "regular" capacitors might 'tune' the decoupling performance over a wide range of frequencies. If this analysis is performed using SPICE circuit type analysis, which ignores the three dimensional full wave nature of PC boards, the high-frequency capacitors appear to provide a lower impedance (and therefore lower the transfer function) across a wide frequency range. The three-dimensional full wave nature of PC board is an important behavior not accounted for in SPICE simulation, and combined with the added inductance of the capacitor attachment, "tuning" the capacitor values does not yield the results initially expected.

To illustrate this limitation, the 0.01 uF capacitors used in the previous section are now combined with 330 pF capacitors to illustrate the effect on the high-frequency performance of S_{21} when two different values of capacitors were used. For this set of experiments, the test PC board is completely populated with alternating .01 uF and 330 pF capacitors (all 99 locations had either 0.01 uF capacitor or a 330 pF capacitor, but not both values).

The results in Figure 8-11 show that at frequencies below about 75 MHz, the S_{21} behavior is very similar to the case with only 0.01 uF capacitors. The S_{21} transfer function is dramatically worse in the 75 – 200 MHz range due to cross resonances (the capacitance of one capacitor resonates with the inductance of the other capacitor and the inductance of the planes between them). There is no noticeable improvement in high-frequency decoupling performance with the addition of the second value of capacitance. In fact, the overall decoupling performance is degraded in a frequency range where much of the typical noise energy exists (50 – 200 MHz).

This example was repeated with a different decoupling capacitor configuration. In this second example, the test board is again fully populated with 0.01 uF capacitors (99 capacitors) and 22 pF capacitors are added directly on top of the SMT 0.01 uF capacitors (connected in parallel). While this mounting strategy is not practical in real world manufacturing, it serves to lower the additional connection inductance for the 22 pF capacitors.

Figure 8-12 shows the transfer function results for various quantities of 22 pF capacitors added onto the 0.01 uF capacitors. The effect on the transfer function is minimal and shows no significant improvement. Given this and the previous example with so-called "high-frequency" decoupling capacitors, there is no improvement, and in some cases the decoupling performance is worsened! Global decoupling capacitors should be all the same value to avoid creating unexpected resonances.

Decoupling Power/Ground Planes / 139

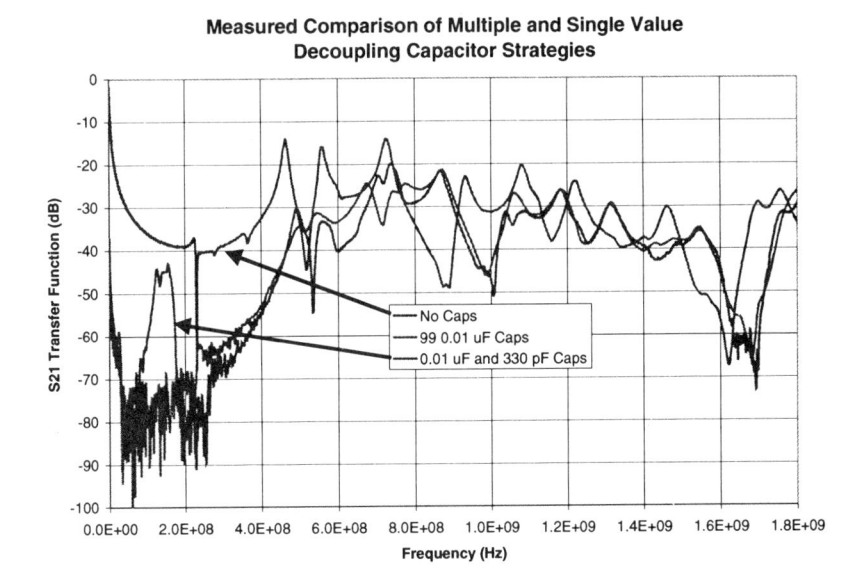

Figure 8-11 Transfer Function with Combination of Decoupling Capacitor Values

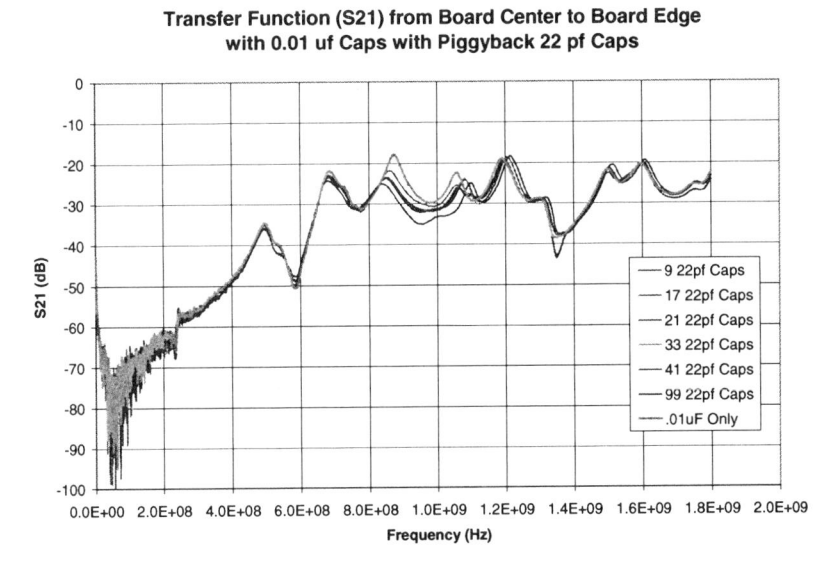

Figure 8-12 Transfer Function with 200 pF Capacitors Added

140 / PCB Design For Real-World EMI Control

8.4.5 Selecting the Value of the Decoupling Capacitors

The actual value of the decoupling capacitors is another topic for confusion. The decoupling performance of global SMT 0805 capacitors with values of 0.01 uF, 0.1 uF, and 0.33 uF on the same test PC board is shown in Figure 8-13. The high-frequency performance of the transfer function is not significantly affected by the value of the decoupling capacitor selected for these capacitor values. Again, the inductance of the capacitor and of its attachment to the board overwhelms the impedance of the capacitor at high frequencies. The optimum solution is to choose an SMT capacitor package size, and then use the ***largest capacitance value available in that package size.***

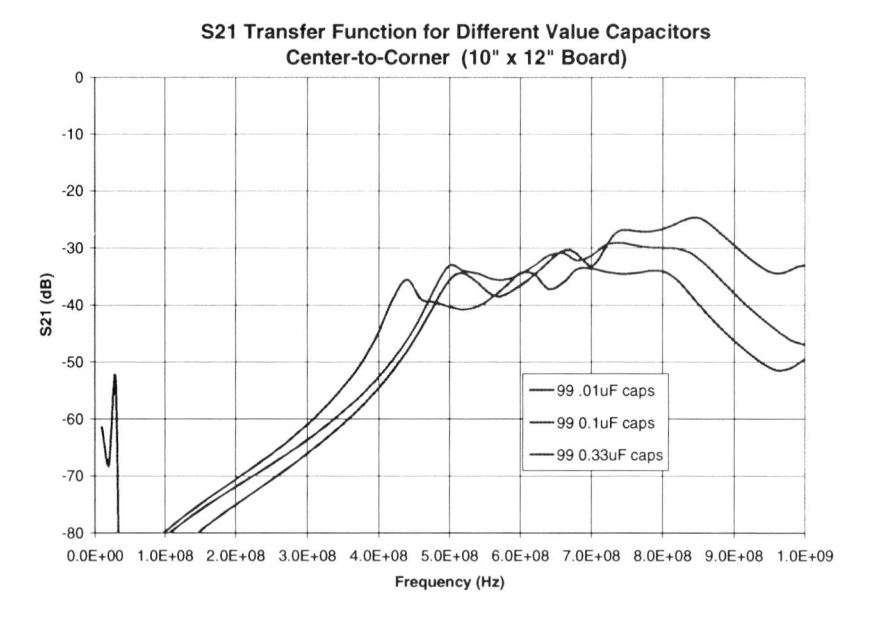

Figure 8-13 Transfer Function for Different Capacitor Values

8.4.6 Perfect Decoupling Capacitors

A perfect decoupling capacitor can be approximated as a simple via between two adjacent parallel planes. While this is impractical when the two planes have different DC voltages, it is valid for planes of the

Decoupling Power/Ground Planes / 141

same DC voltage (e.g. two ground-reference planes) and it serves as a reasonable limit to the performance of decoupling capacitors.

Figure 8-14 shows the transfer function performance of the various global decoupling capacitor values and the results when only a via is present. Again, the natural inductance of the via limits the transfer function at high frequencies. Clearly, the capacitors are about at the limit of performance and the transfer function cannot be reduced. The analysis with "perfect" capacitors shows that the value of the capacitor is not important at high frequencies. As stated in the previous section, the largest capacitance value (for a given package size) is best for low frequencies, and should always be selected for global decoupling capacitors.

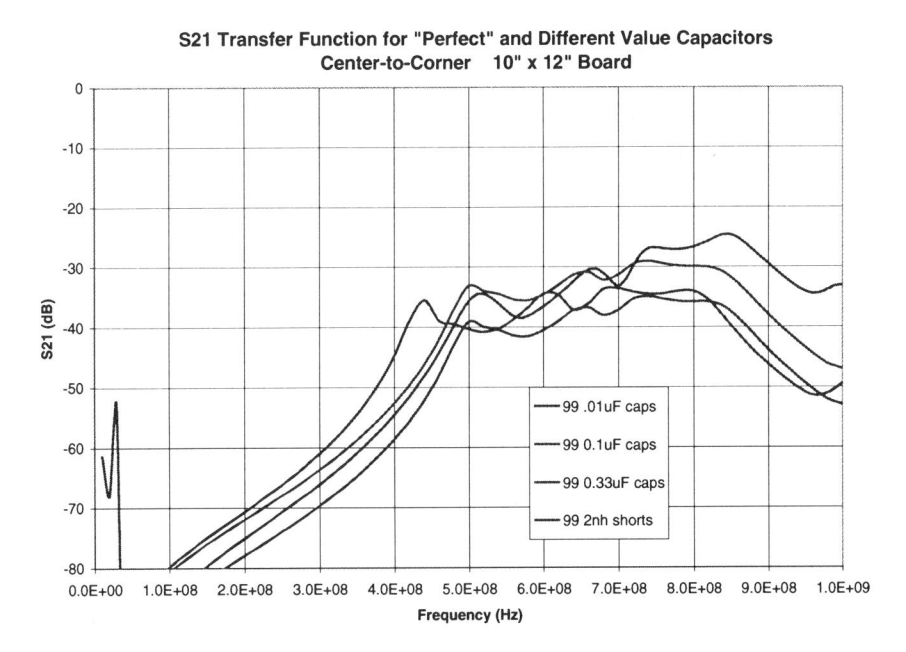

Figure 8-14 Transfer Function with "Perfect" Capacitors

8.4.7 Source Vs Distributed Decoupling

As mentioned earlier, a common debate between EMC and design engineers is whether to place decoupling capacitors close to the source of the 'noise' (that is, near the power and/or ground pins of the

ICs), or to simply distribute them across the entire board. If we consider the purpose of the capacitors, we can easily answer this question. First, capacitors must deliver charge to the IC/ASIC power pins quickly when the device changes from a high impedance to a low impedance state internally. In order to reduce the loop inductance, the capacitor must be positioned as close to the power pins as possible. The second purpose of the decoupling capacitors is to keep the noise between the power and ground-reference planes as low as possible across the board and especially along the edge of the board.

The transfer function, or transfer impedance, can be used to observe the remote effects of adding local decoupling capacitors. In [8.3], a 6" x 9" PC board was tested with global decoupling capacitors spaced on a one-inch grid. Figure 8-15 shows the effect on the transfer impedance when only the global decoupling capacitors were used and when a local decoupling capacitor is added at various distances from the source. There is about a 5 dB improvement in the transfer impedance when the local capacitor is placed extremely close (50 mils) to the source, but the improvement drops to 1-2 dB when the capacitor is placed 300 mils from the source. In most applications, it is unrealistic to place capacitors within 50 mils of all the power pins in high density ASICs. The expected reduction in the EMI noise at a location away from the local decoupling capacitor is minimal, but a capacitor should still be placed close to the power pins for the charge delivery function.

This is consistent with our understanding of the cause of the high-frequency transfer function performance. The resonant nature of the three-dimensional cavity formed by the closely spaced parallel plates dominates the high-frequency performance of the transfer function.

It is helpful to consider the following analogy to help understand how decoupling capacitors work in this resonant environment. Consider a metal shielded room. We place a vertically oriented dipole antenna at some location in the room, and excite the antenna with a high-frequency sine wave. If the frequency is selected so that it is a resonant frequency of the shielded room, standing waves will be created and there will be locations where the electric field is greatest and other locations where it is minimal. Next we place a metal post from floor to ceiling near the source antenna. When observing the electric field throughout the room, we would observe a change in the location of some of the maximum field locations, but the field strengths would not be significantly reduced.

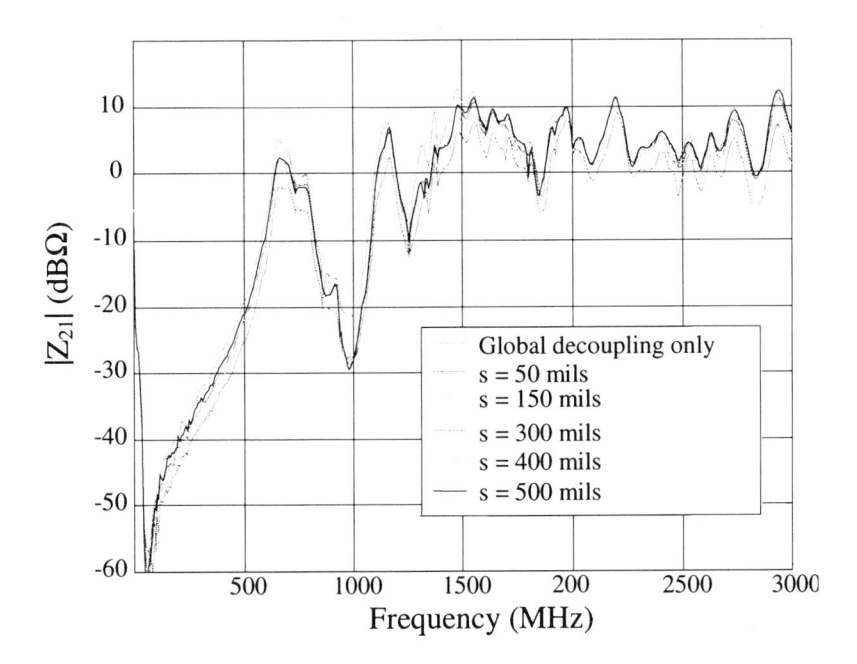

Figure 8-15 Transfer Impedance with Local Decoupling

Next, let's add a number of metal posts throughout the room. Again, we have changed the boundary conditions in the room, and the field distribution will change. If the metal posts are spaced close enough, the standing waves may not be able to fully develop, and the maximum electric fields will be lower than the previous case.

This analogy can be helpful to understand the reason that positioning a decoupling capacitor near the IC power pins does not significantly affect the decoupling transfer function. This does not imply that placing decoupling capacitors near the IC power pins is unnecessary, however. Remember that the capacitors placed near the power pins provide a rapid charge delivery to the IC for functionality considerations and are very important for this task.

While a distributed decoupling strategy appears to be best for an EMC design, signal integrity design considerations still require a decoupling capacitor near a high-speed IC to provide the necessary supply current with as low a series inductance as possible. A combination of both decoupling strategies will meet both EMC and signal integrity requirements.

8.4.8 Buried Capacitance Decoupling

The so-called *buried capacitance* is created by placing the two planes close together, thus increasing the inherent capacitance between the planes. In some cases, the dielectric constant of the material between the two planes is increased, further increasing the inherent capacitance between the planes.

Figure 8-16 shows the transfer function for the same 10" x 12" board with the spacing between the planes decreased and the same dielectric (FR4) used. The transfer function is lowered significantly when the spacing between planes is only 2 mils compared to a separation of 35 mils. When compared to the 35 mil separation case with 99 global 0.01 uF SMT capacitors, the transfer function is reduced at high frequencies, but not at low frequencies. This is the first example we have seen where high frequencies were significantly lowered. A combination of traditional decoupling capacitors and closely spaced planes will give an overall lower transfer function while potentially reducing the total number of capacitors needed.

It is important to understand that closely spaced planes provide decoupling for those planes only. In a PC board where closely spaced planes are used, there are usually a number of different layers, some planes and some signal layers and possibly a number of planes. The transfer function is improved only for the closely spaced planes. Since noise between planes can also be created by high-speed signal via transitions, [8.4] other pairs of planes still need traditional decoupling capacitors. Figure 8-17 shows an example to illustrate this concept.

Decoupling Power/Ground Planes / 145

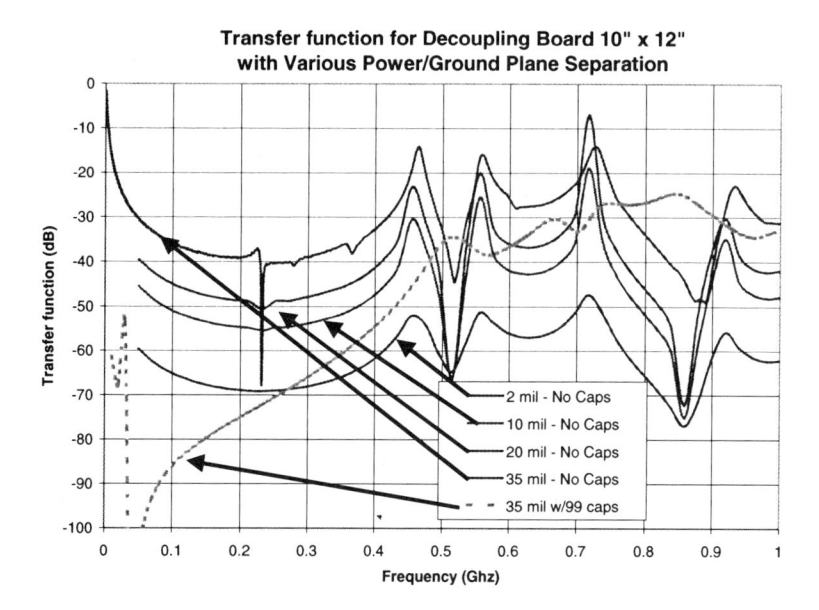

Figure 8-16 Transfer Function with "Buried Capacitance" Decoupling

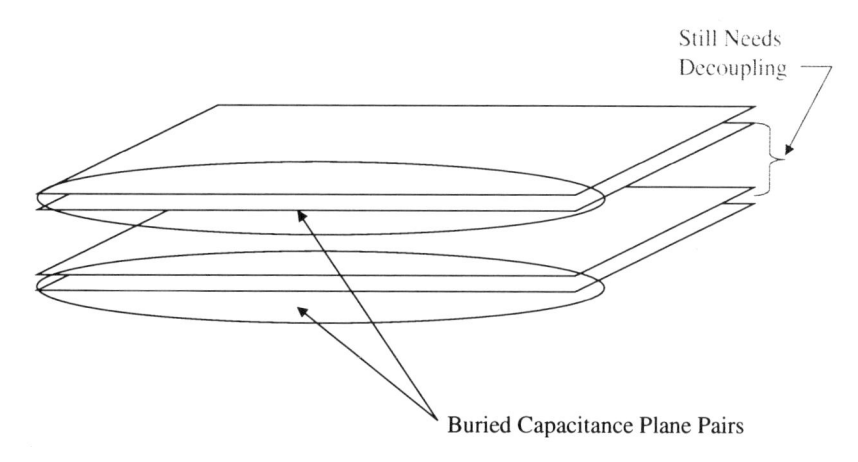

Figure 8-17 Example of Multiple Plane Pairs

8.4.9 Lossy Capacitors

It is commonly accepted that decoupling capacitors should have low equivalent series resistance (ESR). It has been shown that at high frequencies, normal circuit simulation results are not consistent with the measured results, and the physical resonances of the board will dominate at high frequencies. It is therefore necessary to add some amount of loss to dampen the Q factor of these resonances. Experiments with high ESR capacitors (6 - 10 ohms in series with a 0.1 uF capacitor in the same SMT package) showed that lossy capacitors can have a significant effect at high frequencies. Figure 8-18 shows the results from these measurements. Figure 8-19 shows the measured impedance of the SMT part for both the typical low ESR capacitor and the lossy capacitor. The best improvement in the transfer function occurs when approximately 65% of the total capacitors used are lossy and the remaining 35% are regular low ESR capacitors. With this combination of low ESR (regular) capacitors and lossy capacitors, the high-frequency transfer function was lowered as much as 20 dB without significant degradation in the low frequency performance.

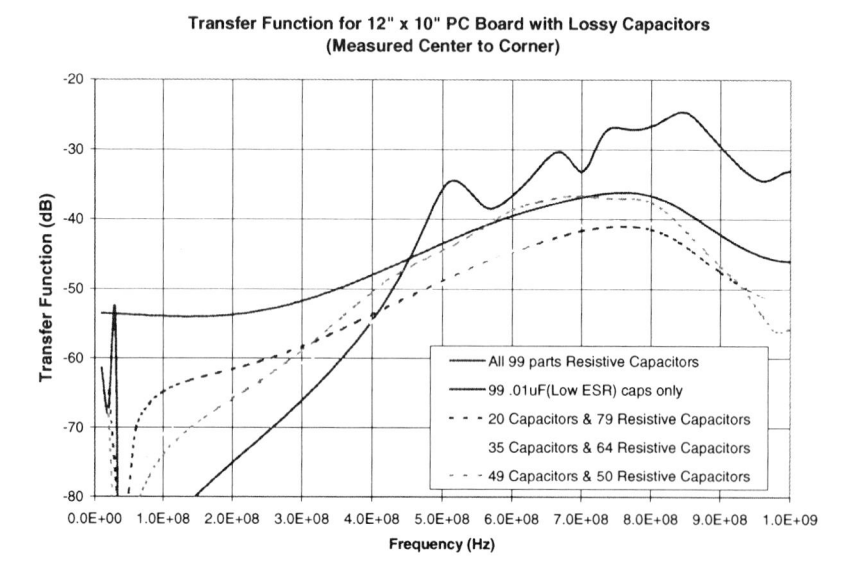

Figure 8-18 Transfer Function with Lossy Decoupling Capacitors

Note in Figure 8-18 that using only lossy capacitors significantly degrades the low frequency performance of the transfer function. A combination of both types of capacitor provides the optimum performance. The low ESR capacitor is a low impedance only at low frequencies above which the natural inductance of the SMT part (and interconnect inductance) become dominant. At frequencies where the low ESR capacitor is no longer effective, the loss of the lossy capacitor lowers the resonance effects.

It is important to remember that the decoupling capacitor serves more than one purpose. While our focus has been EMI noise reduction, another primary function of these capacitors is to provide charge (power) to the IC or ASIC rapidly during device switching. In order to provide the charge quickly, the decoupling capacitor must have a low ESR in order to insure a low RC time constant. When using lossy capacitors, it is important to use low ESR capacitors near the IC/ASICs and to use the lossy decoupling capacitors only for the remote global decoupling capacitors.

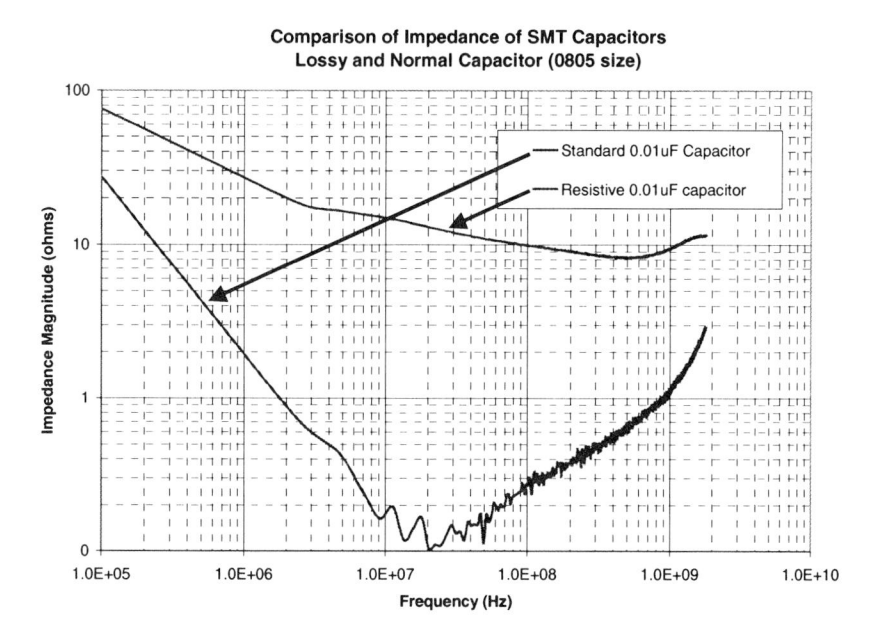

Figure 8-19 Impedance of Lossy Capacitor

Another important purpose of decoupling capacitors is to provide a low impedance return path for current when a high-speed trace changes reference planes on a via. This subject is discussed in Chapter 7. Since these capacitors must provide a low impedance path for the return current, lossy capacitors should not be used for return current path applications.

Lossy capacitors can have a significant effect on the high-frequency performance of the overall decoupling strategy without requiring special board designs (as in the case of distributed capacitance between planes), but they require care in their use. The general approach of simply sprinkling capacitors over the entire board without giving thought to the noise source can cause problems. The purpose of the capacitor must be considered, and the appropriate component used at each location on the board.

8.5 Summary

Decoupling design strategy is important and complex. This chapter has shown how to predict the level of decoupling noise likely from various devices. Using C_{pd} for clock buffers along with the I/O current can give a good estimation of the amount of noise expected. Even if the C_{pd} parameter is not available from the manufacturer, a reasonable estimation can still be made using only the I/O current. This is especially true in cases where there is a large number of output drivers or the I/O current is significant (as in the case of DDR RAM memory).

The impact of decoupling capacitor location, value(s), and the effectiveness of adding so-called "high-frequency" capacitors was also discussed. Global decoupling with a fixed value was shown to be the most effective while also preventing unexpected resonances. In general, once the SMT capacitor package size is selected, the largest capacitance available in that package size should be used. A variety of different capacitor values does not reduce the high-frequency resonance effects, and in some cases can reduce the effectiveness of the decoupling strategy in mid-range frequencies.

Other decoupling strategies can be effective if done correctly. The so-called buried capacitance is really just an increase in distributed capacitance created when the planes are spaced closer than normal or when the dielectric constant of the inter-plane material is increased. This can be effective, but only for the two planes involved, and it does not reduce the need for traditional decoupling capacitors for other pairs of planes in a multilayer PC board stackup.

Lossy decoupling capacitors were shown to reduce the resonance effects at high frequencies by adding loss. This can be very effective in reducing noise between the power and ground-reference planes, but lossy capacitors must be used carefully. A combination of lossy capacitors and regular low ESR capacitors provides the best performance. Care is needed to make sure that capacitors placed near ICs and high-speed signal via transitions are low ESR capacitors.

References

[8.1] Wei Cui, *Modeling and Design of DC Power Bus Interconnects, Segmentation, and Signal Via Transitions in Multi-Layer Printed Circuit Boards using FDTD and a Mixed-Potential Integral Equation Approach with Circuit Extraction*, PhD, May 2001.

[8.2] H. Shi, F. Sha, J. L. Drewniak, T. H. Hubing, and T. P. VanDoren, "An experimental procedure for characterizing interconnects to the DC power bus on a multi-layer printed circuit board," *IEEE Trans. Electromagn. Compat.*, vol. 39, pp. 279-285, November 1997.

[8.3] W. Cui, J. Fan, S. Luan, and J. L. Drewniak, "Modeling shared-via decoupling in a multi-layer structure using the CEMPIE approach," *10th Topical Meeting on Electrical Performance of Electronics Packaging, Cambridge*, Massachusetts, pp. 265-268, October 29-31, 2001.

[8.4] W. Cui, X. Ye, B. Archambeault, D. White, M. Li, and J. L. Drewniak, "Modeling EMI resulting from a signal via transition through power/ground layers," *Proceedings of the 16th Annual Review of Progress in Applied Computational Electromagnetics*, Monterey, CA, pp. 436-443, March 2000.

Chapter 9

EMC Filter Design

9.1 Introduction

Traditional filter design is complex. The maximum pass band ripple allowed, the roll off required, and the minimum loss in the stop band all must be considered. There are various standard filter types with linear phase, faster roll off, etc. Once it is decided which of the standard filter types is desired, the realizablility of the filter must be considered, and after all this, then we can start to assign values to the components of the filter. While these traditional filters are necessary in many applications, such as radio receivers, they are typically more complex than the real-world EMC application requires.

A filter is really a simple voltage or current divider. If the impedance of the filter is considered, its source impedance, and its load impedance, and simple Kirchoff laws are applied, then the filter can be quickly designed and it will meet the needs for most EMC concerns without the need for the traditional filter complexity.

9.2 Filter Design Concepts

The most common EMC filter is one to stop unwanted signals from exiting a shielded enclosure and causing emissions from the external cable. Figure 9-1 shows a simple case where an I/O driver is some distance from the I/O connector and a filter is used to stop unwanted signals from exiting on the cable. These signals may have originated

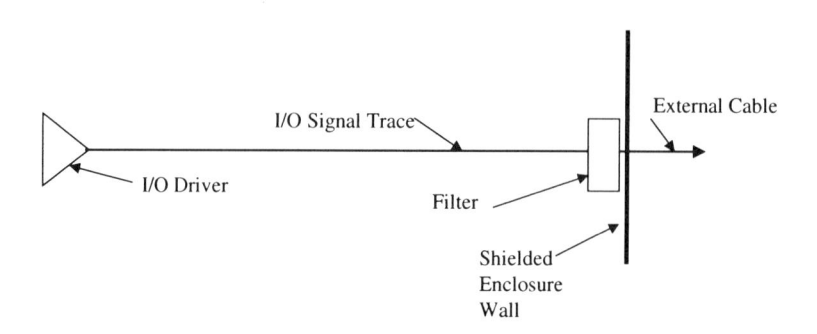

Figure 9-1 I/O Filter Example

from a number of sources, including crosstalk coupling[1] from critical signal traces, I/O driver noise, or from some other general EMI coupling path onto the I/O trace. In order to minimize high-frequency pickup between the filter and the I/O connector, the filter should be placed as close to the I/O connector as possible, within one centimeter for most systems. Additional distance between the I/O connector and the filter components allows high-frequency noise to bypass the filter completely or to couple onto the signal trace between the filter and the I/O connector.

Many I/O signals are relatively low frequency, that is, below 1 MHz. There are some I/O signals that contain much higher frequencies, but these signals typically use shielded cables, and are handled in a different manner as compared to low speed I/O signals using unshielded cables. Regardless of the intentional I/O signal frequency, optimum filter design needs to consider what frequency harmonics are required for proper operation, then the filter should be designed to pass those signals and block all other signals. EMC filters for this application are normally low-pass filters. Figure 9-2 shows the four most common general low pass filter configurations.

Before the effectiveness of these filter configurations can be properly analyzed, the load impedance expected at the I/O connector must be considered. Analysis of the exact impedance is difficult, since EMC emissions testing requires the external cables to be moved

---

[1] As discussed in Section 7.5

in all possible positions to maximize the emissions at all frequencies. Moving the external cables changes the impedance at different frequencies, making the impedance unpredictable. It is generally accepted that the worse case impedance (the external cable's "antenna" radiation impedance) is 100 Ω across the frequency range of interest.

For simple filter design, it is desirable to keep the unwanted signals from reaching the external 100 Ω radiation impedance. This means the series filter components must be selected with impedance much greater than 100 Ω, and parallel components with impedance much less than 100 Ω impedance over the frequency range of interest.

For example, let's assume a 100 KHz intentional I/O signal, and wish to filter all signals from 30 MHz to 1000 MHz. The intentional I/O signal's rise time can be included in the first 7-9 harmonics, so if up to 1 MHz is included in the pass band, the data signal should not be attenuated. A typical rule-of-thumb is to have a factor of 100 between the external wire's radiation resistance and the filter components. This means that at 30 MHz, it is desirable to select the series components with at least 10K Ω and the parallel components with no more than one ohm impedance. For a low pass filter, the parallel components will be capacitors and the series components will be inductors or ferrite beads[2]. While it is not practical to require 10K Ω from ferrite beads, 1000 Ω ferrite beads are available and a one-ohm requirement for the parallel components is achievable. The impedance of these components will be discussed in later sections, but for now, the general impedance of capacitors and inductors is given by (9.1) and (9.2).

---

[2] Ferrite beads are more commonly used because they have a wider effective frequency range and include loss.

(a) Shunt-Series Configuration

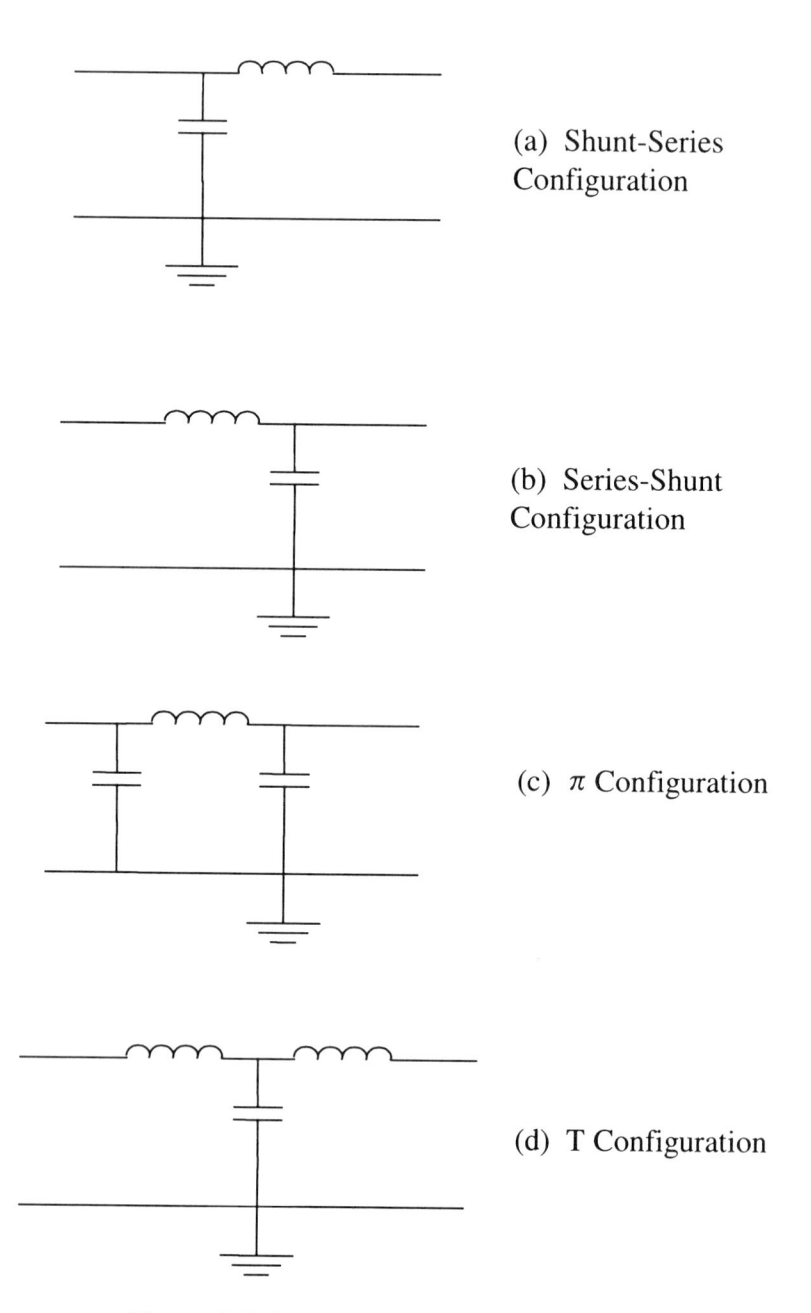

(b) Series-Shunt Configuration

(c) π Configuration

(d) T Configuration

Figure 9-2 Low Pass Filter Configurations

$$\text{Magnitude } X_C = \frac{1}{2\pi fC} \tag{9.1}$$

$$\text{Magnitude } X_L = 2\pi fL \tag{9.2}$$

where:
X_C = impedance of a Capacitor,
C = capacitance,
f = frequency,
X_L = impedance of an inductor,
and L = inductance.

9.3 Filter Configurations

There is some debate about which filter configuration is best. There are a number of considerations that affect this decision. Often, the number of components must be minimized to reduce cost, and/or board real estate used. In this case, a two-component filter, or even a one-component filter is desired. Another consideration is whether the filter is intended to reduce emissions or susceptibility. Two-component filter configurations will be discussed first, since their analysis illustrates filter operation most clearly.

9.3.1 Two-Component Filter Configurations

Figure 9-2a shows a simple filter configuration with a capacitor and an inductor (or ferrite bead). In order to analyze the effectiveness of this configuration, this filter will be inserted into the emissions example in Figure 9-1, and simple current node analysis used.

Consider an unwanted current on the trace traveling towards the I/O connector. Figure 9-3 shows a schematic of this circuit with the external cable's radiation impedance added. At Node #1, the current will "see" a low impedance through the capacitor back to its source relative to the high impedance of the inductor/ferrite. Therefore, very little current will travel through the inductor/ferrite and result in a

156 / PCB Design For Real-World EMI Control

common-mode current on the cable causing external emissions. This is an effective filter.

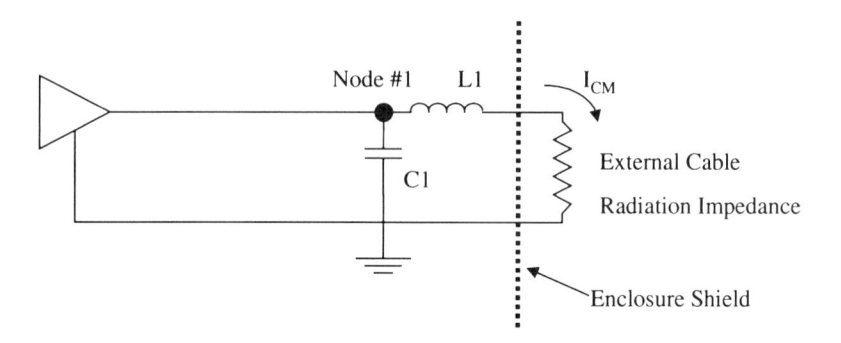

Figure 9-3 Low Pass Filter Configuration (a)

Now consider the filter configuration in Figure 9-2b. Figure 9-4 shows the equivalent schematic. Also, consider the same unwanted current on the trace traveling towards the I/O connector. The current will see a high impedance in the inductor/ferrite, but once it arrives at Node #1, it sees a low impedance in the capacitor across the external cable's radiation impedance. While 100 ohms is often used as a worst-case cable antenna radiation impedance, remember this can be nearly any impedance, depending on the cable position. Due to this uncertainty it can no longer be certain the capacitor will have a low impedance relative to the external cable radiation impedance. The capacitor will be useful at some frequencies, and not useful at other frequencies. Since the external cable's radiation resistance is impossible to know in advance, it can only be assumed that the external cable's radiation impedance will be low at a frequency where there is an unwanted emission and the filter will not be effective. The inductor/ferrite is providing all the filtering that can be depended on in this filter, and the capacitor might as well not be used in this configuration for emissions control.

If the design goal is changed from emissions control to susceptibility control, then the source of the unwanted signal changes from the internal I/O driver to the external cable. The load to be considered is the load presented by the I/O driver (or receiver), which is typically low at high frequencies. The analysis now resembles the previous case (in Figure 9-3) where the capacitor presents a low impedance relative to the high impedance of the inductor/ferrite. This filter configuration is effective for susceptibility concerns.

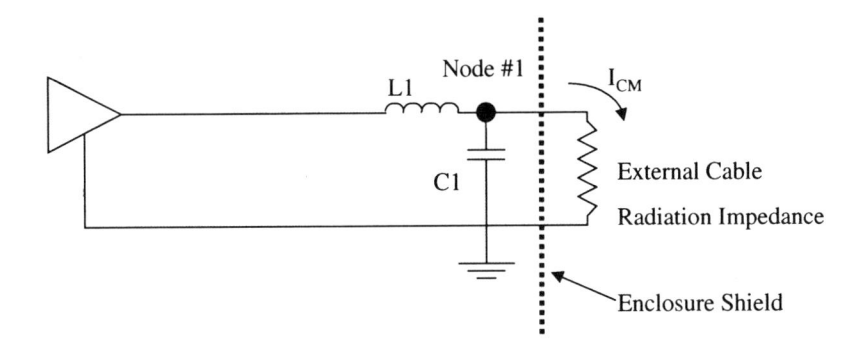

Figure 9-4 Low Pass Filter Configuration (b)

9.3.2 Reference Connection for Two-Component Filters

In the previous section, the position of the capacitor (shunt element) and the inductor/ferrite (series element) for optimum low-pass filtering were discussed. Section 7.4 discussed placing splits in the ground-reference plane to isolate the 'ground' pin of I/O connectors from the spreading return currents from high-speed traces. Which side of this split that the capacitor is connected to reference will have a major impact on the filter's performance.

Figures 9-5 through 9-8 shows the four possible connection configurations with two ground-reference planes connected through a ferrite bead (to provide a return path for the intentional low-speed I/O signals). Figure 9-5 shows the shunt capacitor connected to the 'digital' (or noisy) ground-reference. This is the optimum connection, since it allows high-frequency signals that are on the I/O line[3] to be returned to their source with a low impedance path. The capacitor and inductor/ferrite configuration are optimized to reduce signals from exiting the enclosure as discussed in the previous section.

---

[3] These high-frequency signals may have originated with the I/O driver as unintentional signals, or they could have been coupled onto the I/O line from another source.

158 / PCB Design For Real-World EMI Control

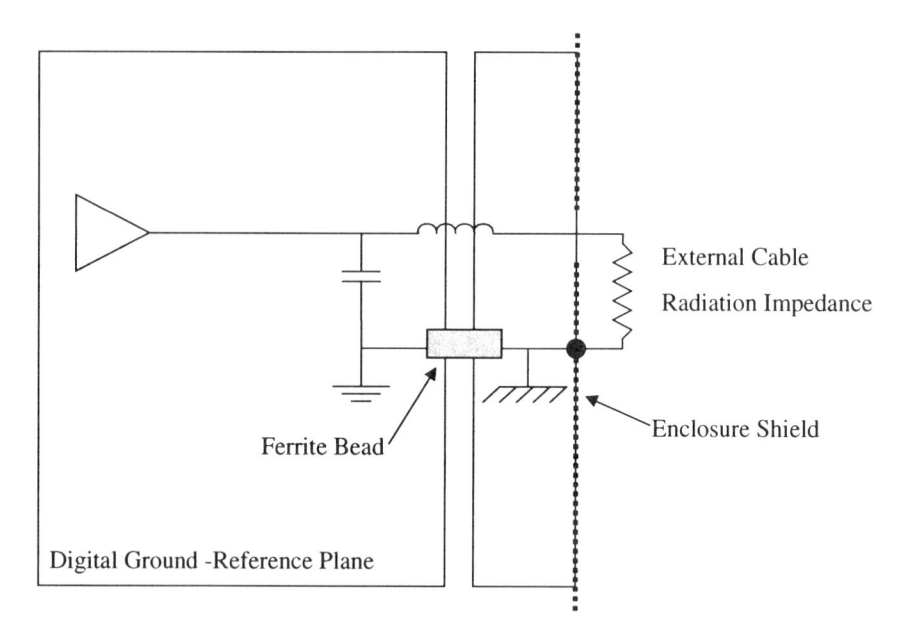

Figure 9-5 Filter Connection Strategy #1

Figure 9-6 shows the same capacitor and inductor/ferrite configuration with the capacitor connected to the chassis reference. This connection would require the high-speed noise signals to pass through the ferrite bead between references. Since a ferrite bead typically has a high impedance to high-frequency signals, this is not a low impedance path back to the noise signal's source and will not be a good filter configuration. This connection strategy should be avoided.

Figure 9-7 uses the opposite capacitor and inductor/ferrite configuration with the capacitor connected to the 'digital' ground-reference. The previous section explained why this capacitor and inductor/ferrite configuration is not the optimal design for emission control. More alarming is the effect on susceptibility for the system. This connection strategy allows any signals from the outside of the enclosure (i.e. ESD pulses, RF susceptibility signals, EMP, etc.) to travel through the low impedance of the capacitor to the digital ground-reference. This will likely cause system susceptibility problems. This connection strategy should be avoided. In addition to being a poor emissions filter, this filter configuration allows external noise (like ESD pulses) to be coupled from the chassis through the capacitor to the data lines. This coupling will lead to likely data errors. Again, this filter configuration should be avoided.

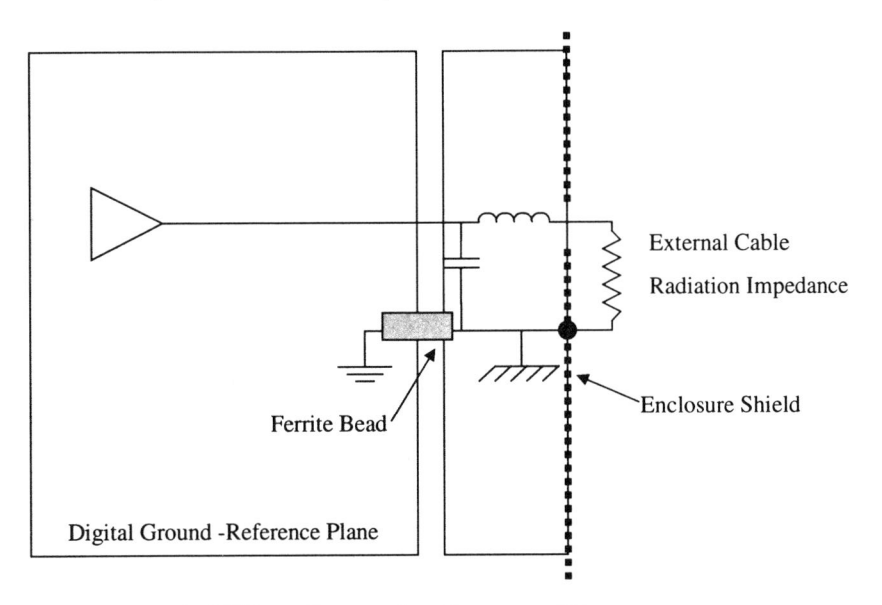

Figure 9-6 Filter Connection Strategy #2

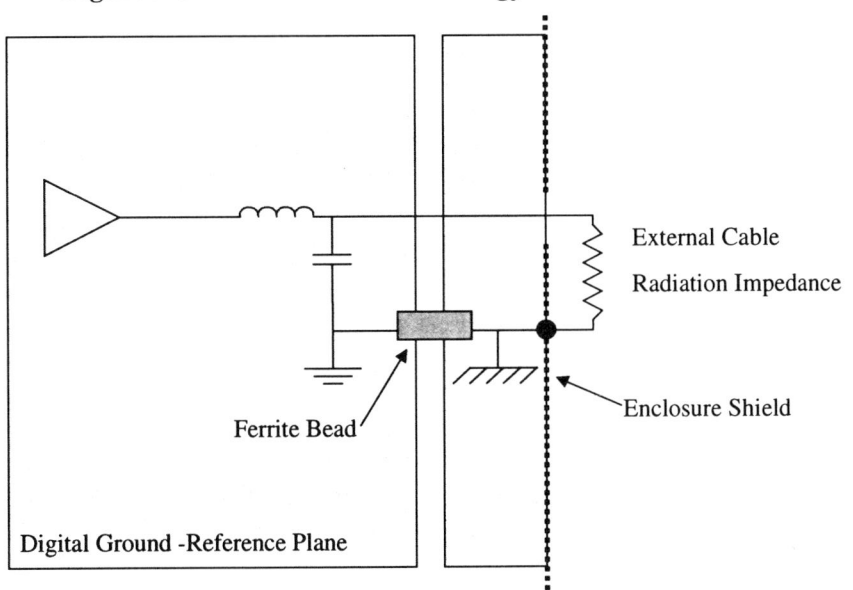

Figure 9-7 Filter Connection Strategy #3

Figure 9-8 shows the same capacitor and inductor/ferrite configuration as the previous example, except the capacitor is now connected to the chassis reference. The capacitor will still not provide a low impedance return current path for internal noise signals trying to exit the system because of the ferrite bead between the two ground-reference planes. For susceptibility concerns, however, the capacitor does provide a low impedance path from the I/O line to the chassis, and will help block these external signals from entering the system. This connection strategy works quite well for susceptibility concerns, but is not effective for emissions concerns. A combination of the connection strategy in Figure 9-5 and figure 9-8 (resulting in a three component filter) would result in optimum performance for both emissions and susceptibility

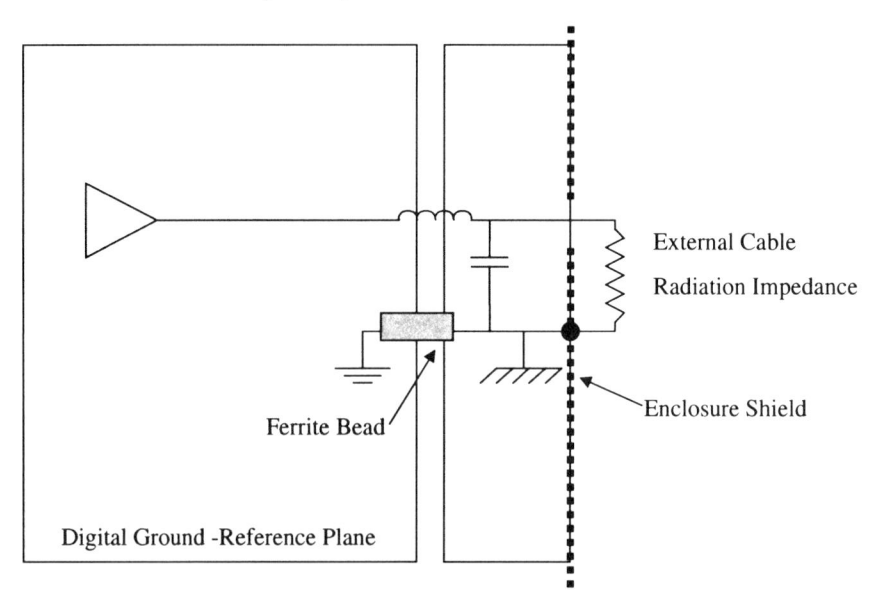

Figure 9-8 Filter Connection Strategy #4

9.3.3 Three Component Filter Configurations

A similar analysis can be performed for the π and T filter configurations in Figure 9-2c and Figure 9-2d, respectively. Figure 9-9 shows the equivalent schematic for the π-filter configuration. Using current node analysis, the unwanted emissions current at Node #1 will see a low impedance path through C1 relative to the high impedance of the inductor/ferrite and be effectively blocked from the external cable radiation impedance. In a similar fashion, any unwanted external signals (susceptibility) will see a low impedance in C2 relative to the high impedance of the inductor/ferrite and be effectively blocked. This means this π-filter configuration is effective for both emissions and susceptibility design concerns.

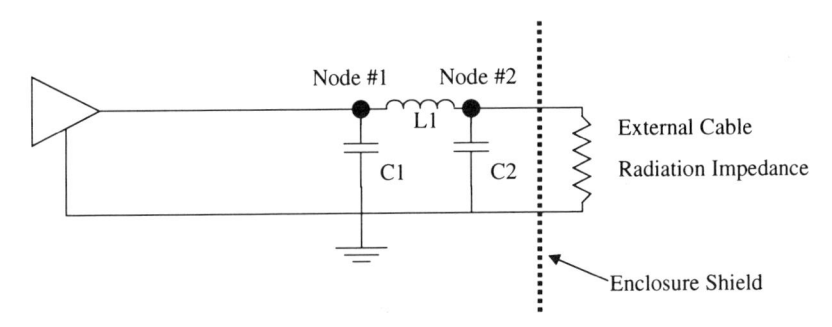

Figure 9-9 Low Pass π-Filter Configuration

In a similar fashion to Section 9.3.2, the connection strategy for the capacitor(s) must be considered. In the case of the π-filter, connecting one capacitor to the digital ground-reference allows the internally created high-frequency noise currents to return to their source (as in Figure 9-5). Connecting the other capacitor to the chassis reference allows external susceptibility signals to return to the chassis without entering the enclosure. This connection strategy is illustrated in Figure 9-10. This is the optimum configuration and should be used whenever possible.

The T-filter analysis is similar. Figure 9-11 shows the equivalent schematic drawing. Using current node analysis similar to that for the previous filter configurations, the unwanted emissions current at Node #1 will see a low impedance path through C1 relative to the high impedance of the inductor/ferrite L2, and be effectively blocked from the external cable radiation impedance. In a similar fashion, unwanted external signals (susceptibility) will see a low impedance in C1 relative to the high impedance of the inductor/ferrite L1, and be effectively blocked. So this means this T-filter configuration is effective for both emissions and susceptibility design concerns.

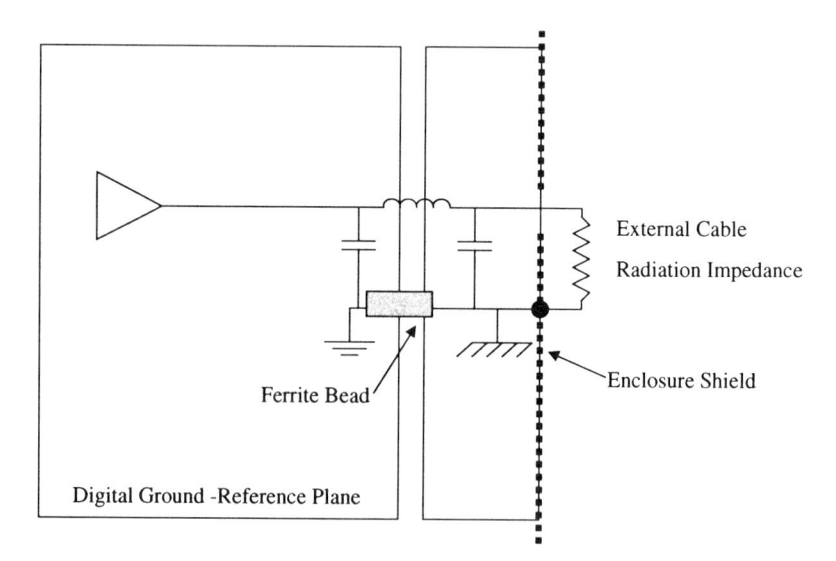

Figure 9-10 Filter Connection Strategy for π-Filter

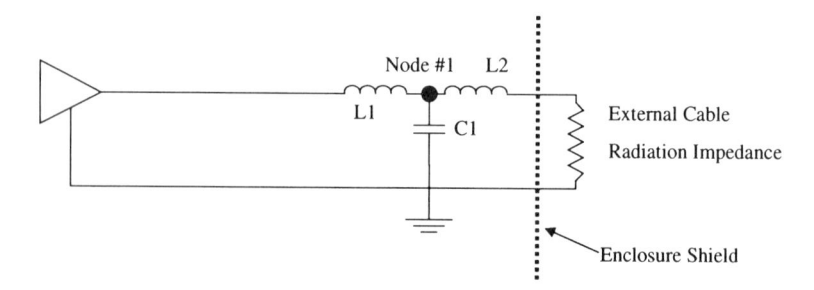

Figure 9-11 Low Pass T-filter Configuration

The T-filter configuration only allows one capacitor connection. If the capacitor is connected to the digital ground-reference, then the filter would be effectively like the filter in Figure 9-7 for emissions and would not be the most effective filter. Conversely, if the capacitor is connected to the chassis reference, the filter configuration resembles Figure 9-6 and is not very effective for emissions or susceptibility. If the load and source impedances are very low, the high impedance provided by the T-filter can be very effective. Since most EMC applications do not have low source and load impedances, the T-filter configuration is not as desirable as the π-filter configuration for EMC applications.

9.3.4 Single Component Filter Configurations

Often a single component must do the filtering due to cost and space constraints on the PC board. A similar current node analysis to those previously discussed will show the most effective single component for this filtering task. If the inductor/ferrite is removed from Figure 9-3, the capacitor will be placed in parallel with the external cable's antenna radiation impedance. When the external cable is in a position so that the external impedance is low, the capacitor will not shunt the high-frequency currents away from the external cable and the capacitor will not be effective as a filter. If the single component filter is an inductor/ferrite, however, then the current will see a high impedance path and be reduced. Therefore, when only a single component is used for a filter, the series inductor/ferrite is more effective than the parallel capacitor. This will become even more apparent when the capacitor's internal inductance and the connection inductance are included in the analysis.

9.4 Non-Ideal Components and the Impact on Filters

In the previous sections the performance of a filter was analyzed using ideal components: a capacitor with only capacitance, and an inductor with only inductance. Real-world components have both inductance and capacitance. The parasitic portion of the component will tend to limit the frequency range where the component is effective.

9.4.1 Non-Ideal Capacitors

When using a simple ideal formula as in (9.1) to find impedance of a capacitor, the true impedance can be underestimated. At high frequencies (above about 100 MHz) the parasitic elements of the capacitor have a significant effect on the impedance of the capacitor. Since a capacitor consists of two or more parallel metal plates separated by a dielectric, the metal plates of the capacitor have inductance. In addition, the connection pads, vias, etc. add inductance. Figure 9-12 shows the impedance of a single surface mounted technology (SMT) 0.01 uF capacitor. Note that the impedance does decrease as frequency increases (and as Equation 9.1 predicts) only to about 60 MHz, and then the impedance actually *increases* with frequency. The parasitic inductance of the SMT capacitor dominates the impedance of the component, and the component no longer acts like a capacitor. Figure 9-13 shows a simple schematic representation of the capacitor with its parasitic elements.

This parasitic inductance is increased when there is more metal in the component. For example, low-speed applications often use capacitors with wire leads. Figure 9-14 shows the difference in the impedance for two capacitor values when the SMT part and the capacitor with leads are measured. Notice that the resonant frequency is greatly lowered when the capacitor has leads, and the impedance at high frequencies is 20 – 24 dB greater compared to the surface mount capacitor.

Even when using SMT capacitors, another important consideration is the connection impedance of the traces and vias. Each via typically adds approximately 0.75 nH of inductance. The length of trace connecting the SMT capacitor to the via is also important. An additional trace length of two centimeters will add significant inductance [9.1]. Clearly, the capacitor will not be very effective at high frequencies once all the connection parasitic inductances are included if these inductances are not minimized. In fact, designing and implementing a filter on a multilayer PC board that is effective beyond several hundred megahertz is very difficult.

EMI Filter Design / 165

Measured Impedance of .01 uf Capacitor

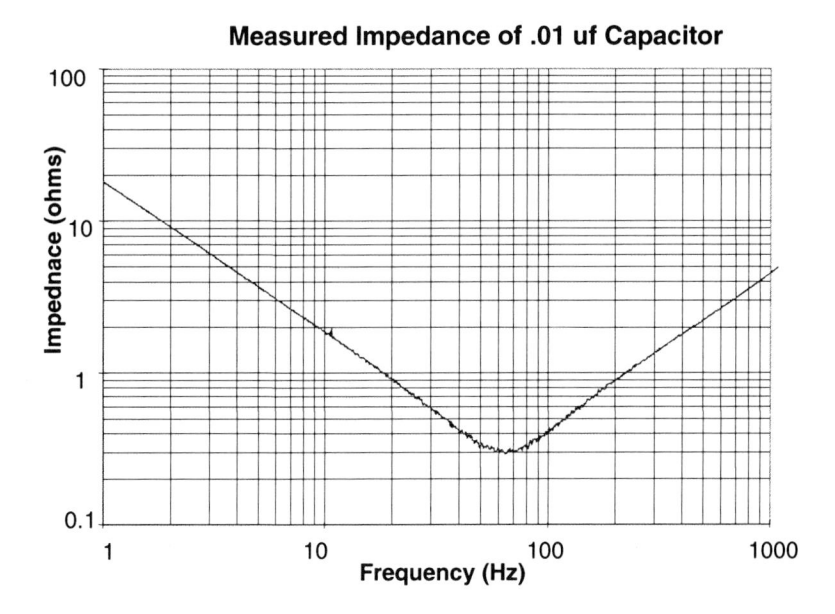

Figure 9-12 Impedance of SMT 0.01 uF Capacitor

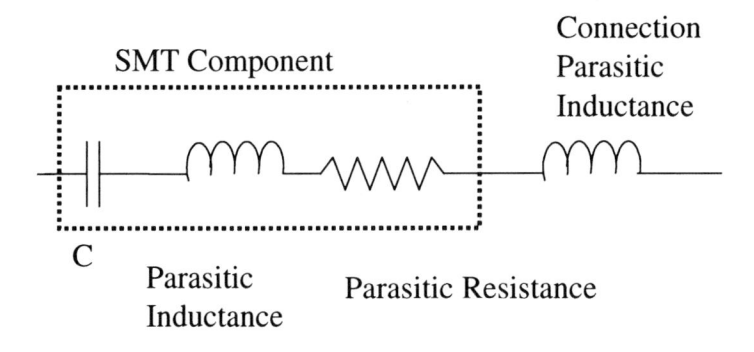

Figure 9-13 Equivalent Circuit of SMT 0.01 uF Capacitor

166 / PCB Design For Real-World EMI Control

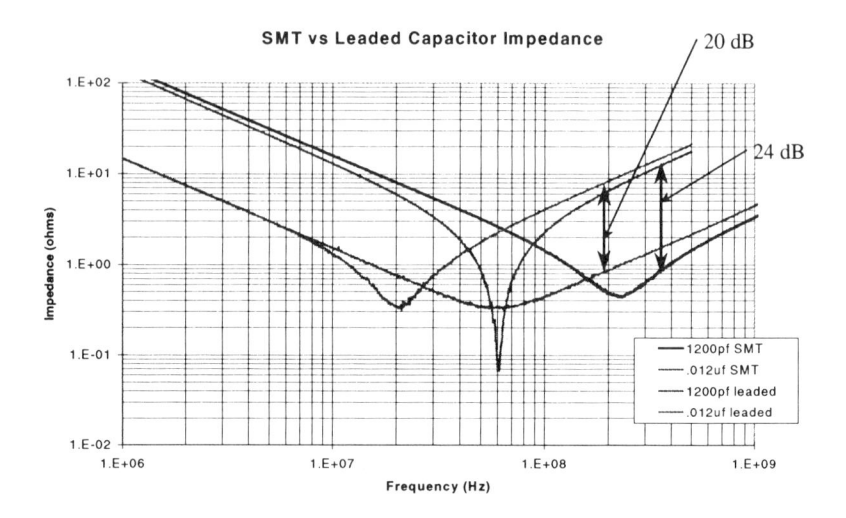

Figure 9-14 Impedance of Leaded Capacitors

9.4.2 Non-Ideal Ferrite Beads

Ferrite beads are a widely used and effective SMT filter component. They resemble an inductor since their impedance increases with frequency (until the parasitic capacitance becomes dominant), and they usually have a low Q-factor so the loss they provide to the filter has a wide bandwidth. Figures 9-15 and 9-16 show the impedance of some typical SMT ferrite beads. Ferrite beads are usually specified by their impedance at 100 MHz. Because of their frequency dependence, care should be taken when using this method of specifying ferrite beads. For example, two of the beads in Figure 9-15 have an impedance of 80 ohms at 100 MHz, but their impedance above 100 MHz is quite different from each other.

The impedance does not always continue to increase as the frequency increases above 100 MHz. Figure 9-16 shows an example of a few high impedance SMT ferrite beads. Note that the impedance is at its peak well below 100 MHz.

EMI Filter Design / 167

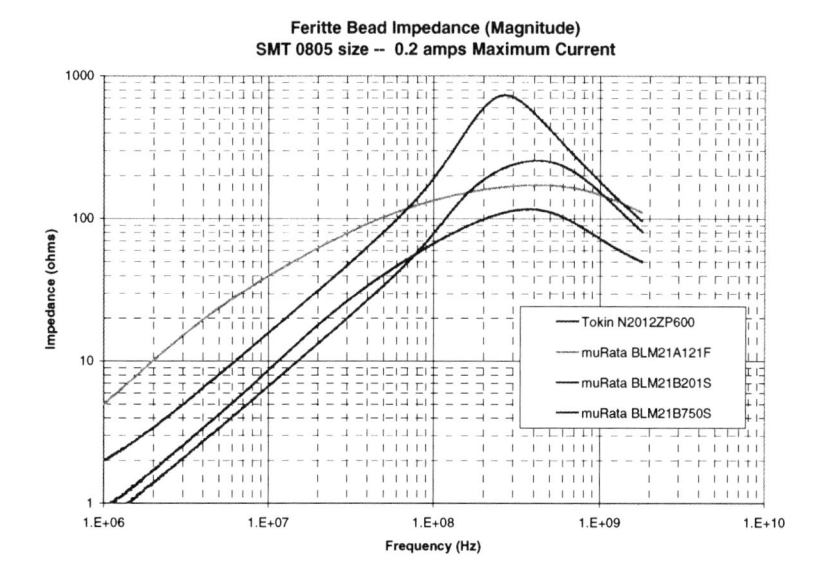

Figure 9-15 Ferrite Bead Impedance Example #1

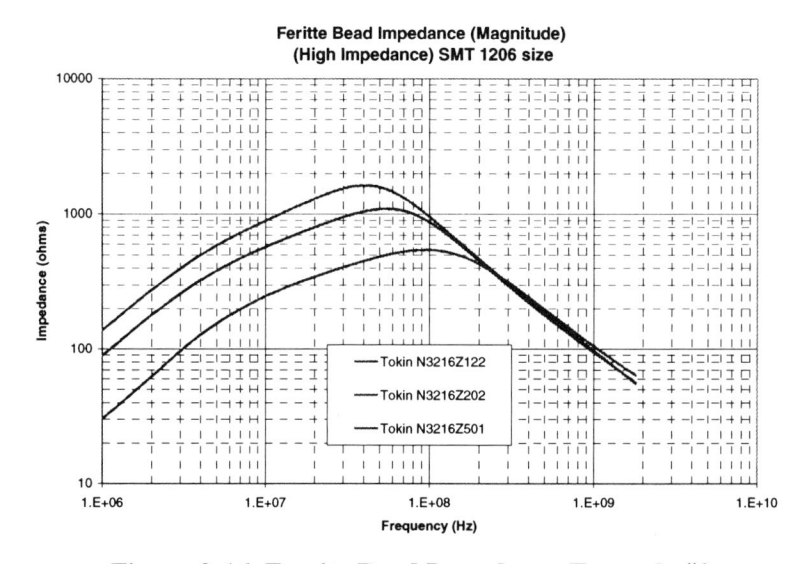

Figure 9-16 Ferrite Bead Impedance Example #2

9.4.3 Non-Ideal Zero Ohm Resistors

Zero ohm resistors are widely used in many designs as 'place holders'. That is, instead of using a ferrite bead, a "zero ohm" resistor might be installed because it is less expensive than a ferrite bead. Note that this resistor is not truly zero ohms, but instead is typically 0.05 ohms[4]. If emissions laboratory testing indicates the need for more filtering, the "zero ohm" resistor can be replaced by a ferrite bead (of the same physical size) without the need to re-layout the PC board.

This technique is quite effective, since the "zero ohm" resistor is actually a small inductor (a zero ohm inductance) as a result of the interconnect and component parasitics and therefore provides a small amount of series filtering. The same general values described previously in Section 9.4.2 for the parasitic inductance of the component, the vias and the connection traces should be used for the "zero ohm" resistor.

9.5 Common-Mode Filters

Many I/O data signals use differential signaling. USB and Ethernet are examples of differential signaling I/O data. Typically, the data is passed though an isolation transformer converting it from common-mode data to differential-mode data. At this point, a common-mode filter should be used to remove all common-mode components of the signal and is often built into the isolation transformer. This common-mode filter is especially important since differential data signaling often uses unshielded cables. Any common-mode data on the external unshielded cables will cause emissions.

Once the differential signal has gone through the common-mode filter, there is no longer any need to have a ground-reference plane beneath the data traces on the PC board as shown in Figure 9-17. The ground-reference plane is only used for common-mode signals. True differential signals do not use this ground-reference. If the ground-reference plane is present, RF noise currents on the ground-reference

---

[4] A true "zero Ohm" resistor would be an expensive precision component.

plane can couple onto the differential signal traces and effectively bypass the common-mode filter. All power planes should also be removed from this area of the PC board for the same reason.

Some external I/O connections require both differential and single-ended signals (e.g. SCSI). These types of I/O require careful analysis of which signals are differential and which are not differential. Non-differential signals should be treated as described previously in this chapter, including treating the "ground" signal return as a signal with a filter, etc. Differential signals should be treated as described above. A schematic example of a mixed I/O connection is shown in Figure 9-18.

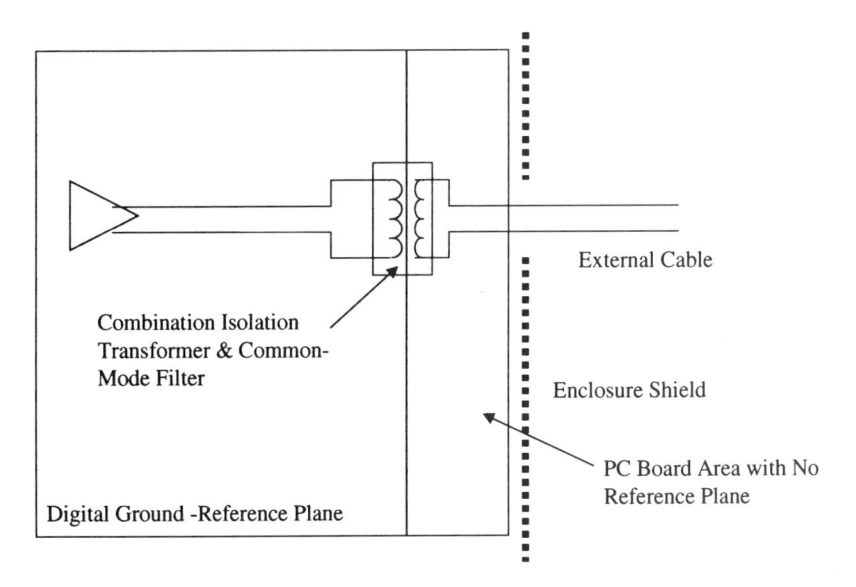

Figure 9-17 Differential I/O Line with Common-Mode Filter

9.6 Summary

Filter design for EMC applications does not usually need to include the complexity of traditional filter design. Using simple current node analysis, the filter components can be easily selected for the case desired. The impedance of the shunt components should be much lower than the impedance of the external I/O cable's antenna radiation impedance. Series elements should have much higher impedance than the external I/O cable's antenna radiation impedance.

Parasitic inductance greatly limits the minimum impedance of a shunt capacitor. The parasitic capacitance of inductors/ferrite is less important. The effect of the parasitic components should be included in the filter design for a realistic evaluation of the frequency range where the filter will be effective.

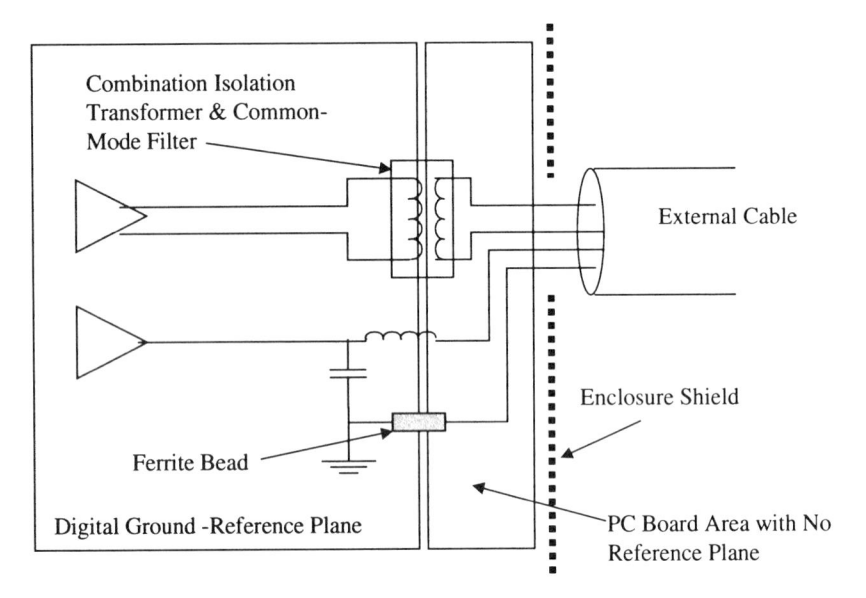

Figure 9-18 Example of I/O Cable with Differential and Single-Ended Signals

References

[9.1] H. Shi, F. Sha, J. L. Drewniak, T. H. Hubing, and T. P. VanDoren, "An experimental procedure for characterizing interconnects to the DC power bus on a multi-layer printed circuit board," *IEEE Trans. Electromagn. Compat.*, vol. 39, pp. 279-285, November 1997.

Chapter 10

Using Signal Integrity Tools for EMC Analysis

10.1 Introduction

Most high-speed printed circuit (PC) boards undergo some degree of signal integrity analysis by engineers using one or more commercial software tools. Engineers analyze the trace layout on the board to ensure the voltage waveform at the receiver meets the required specification for proper operation. Termination resistor values are changed, or even more extreme changes are made, so that the proper voltage waveform arrives at the receiver. Once the voltage waveform is acceptable, the analysis is complete. This can result in a wide range of termination resistor values being used on designs. The value of the termination resistor is not always optimized; as long as it works, it's acceptable. The value of the termination resistor can have an enormous effect on the intentional current on the trace, however.

From an EMC emissions point of view, the voltage waveform is not a concern, but the current waveform is very important. "Voltage" does not radiate directly, but "current" does! It is therefore useful to analyze the currents on a trace as well as the voltage waveforms. Unfortunately, only a few commercial software tools allow this analysis of the intentional currents. The value in performing a little extra analysis far outweighs the time and cost. In fact, some of the commercial tools which can provide current analysis are less expensive than those that do not include this analysis.

This chapter will give a few examples of using a commercial signal integrity tool to find the intentional currents on traces, and more importantly, to find the frequency spectrum of those currents. Changing the termination resistor value or making other termination changes can have a significant effect on the high-frequency harmonic content of the intentional currents. Often, the higher harmonics are not required to create an acceptable voltage waveform, and will only cause emissions problems through some other means (as described in Chapters 6 and 7). The bottom line is that if the current is never created, it cannot become a common-mode current that will cause emissions problems somehow, somewhere in the system.

10.2 Intentional Current Spectrum

The primary cause of EMI emissions is the so-called "common-mode" currents. Basically, common-mode currents are currents that exist at locations where they were never intended. The common-mode current can couple onto a nearby I/O cable or other conductor leaving the shielded enclosure, and then cause emissions.

Common-mode currents can be caused by a number of different sub-optimal design practices, as discussed in detail in previous chapters. The intent of the traces on a PC board is for all the return current to flow directly beneath the trace in the reference plane (usually the power plane or the ground plane). As discussed in Chapter 6 and Chapter 7, not all the return current can flow directly under the signal trace,. The return currents will spread out over the entire plane, trying to reduce the inductance in the return path to the lowest possible level. While most of the return current is under the trace, not all of it is contained there, resulting in currents located in places where they were never intended to be.

Often, the board layout design is not optimal for high-speed signals. For example, if a high-speed clock trace is routed over a split in the reference plane (as when the power plane is split to allow more than one DC supply voltage), the return currents must find some other path to return to the source. Even if a capacitor is placed across the split near the crossing, the added inductance of the capacitor, the necessary vias, pads, etc., will ensure that the high-frequency components of the return current will not be close to the signal trace.

Another common problem occurs when a high-frequency signal trace is routed through a via and changes reference planes. The return current must cross from one plane to the other (possibly through a decoupling capacitor, with its vias, extra inductance, etc.), and often flows in an unpredictable path in order to return to the source.

While the causes of common-mode currents are many, diverse, and often hard to predict, it is true that ALL common-mode currents originate from an *intentional* current. That is, somewhere on the PC board, an intentional signal created the common-mode current unintentionally. It is therefore worthwhile to make sure the intentional signals are controlled so that only the required harmonics exist and to eliminate the unnecessary harmonics. It is more costly to add filtering to an I/O port to stop a high-frequency harmonic from exiting the shielded enclosure when the original signal source did not even need that harmonic for functionality.

A personal computer PC board was chosen to illustrate this concept. A clock net that runs at 133 MHz was selected for analysis. The appropriate driver and receiver IBIS models were used to characterize the driver and receiver. A source series resistor termination scheme was used. The 'default' termination resistor value on this net was 22 ohms.

The voltage waveform at the receiver (for 5 volt logic) was analyzed for termination resistor values from 10 ohms to 39 ohms (typical range). Figure 10-1 shows the effect of changing the termination resistor's value on the voltage waveform. While some amount of pulse amplitude reduction and rise time lengthening occurred as the resistor value was increased, signal integrity requirements would often consider any of the waveforms shown as sufficient to insure proper operation of the system.

Since this analysis was aimed at reducing possible emissions, the current on the trace at the receiver was also analyzed. Figure 10-2 shows the current waveforms for different termination resistor values. It is immediately obvious that the 10 Ω resistor allows much more current to flow than the other values.

174 / PCB Design For Real-World EMI Control

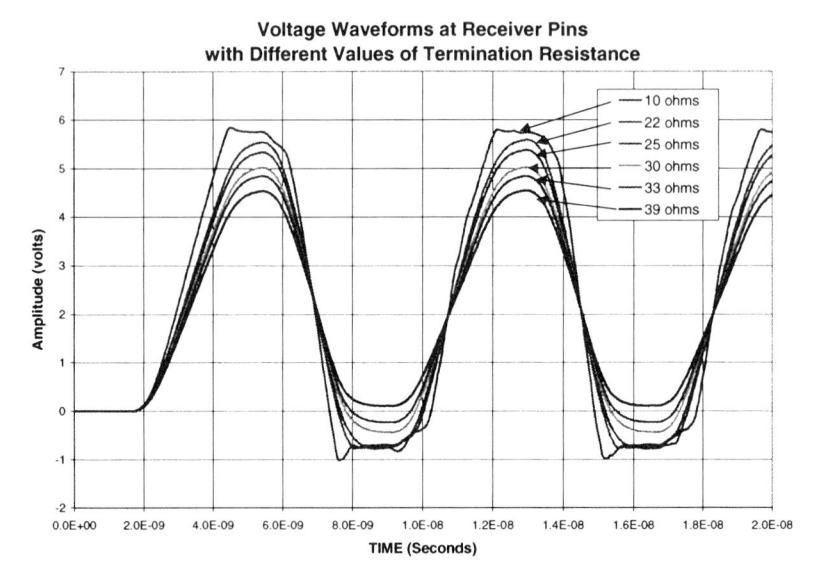

Figure 10-1 Voltage Waveform for Various Termination Resistor Values

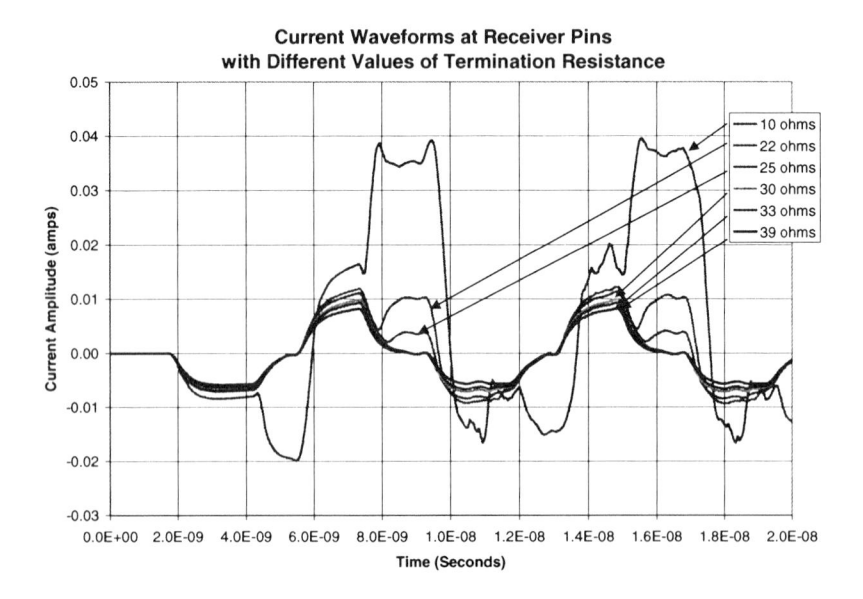

Figure 10-2 Current Waveform for Various Termination Resistor Values

Further analysis shows that the values of 22 and 25 Ω also have extra 'features' that are not present for larger resistor values.

While this voltage and current waveform analysis is useful, it does not really address the amount of reduction of high-frequency harmonics (the most common cause of emissions problems). Fourier transforms of the time domain waveforms were performed to obtain the frequency domain spectrum. These are shown in Figure 10-3 and Figure 10-4 for each of the different termination resistor values (for the frequency ranges of 100 – 1000 MHz and 1000 – 2000 MHz, respectively). The results show a large variation of current amplitude at each harmonic frequency. It can be seen by further analysis that for each harmonic frequency, the amplitude of the current goes down as the resistor value is changed from 10 Ω to 30 Ω, but further increase of the resistor value does not significantly lower the current amplitude at a given harmonic frequency.

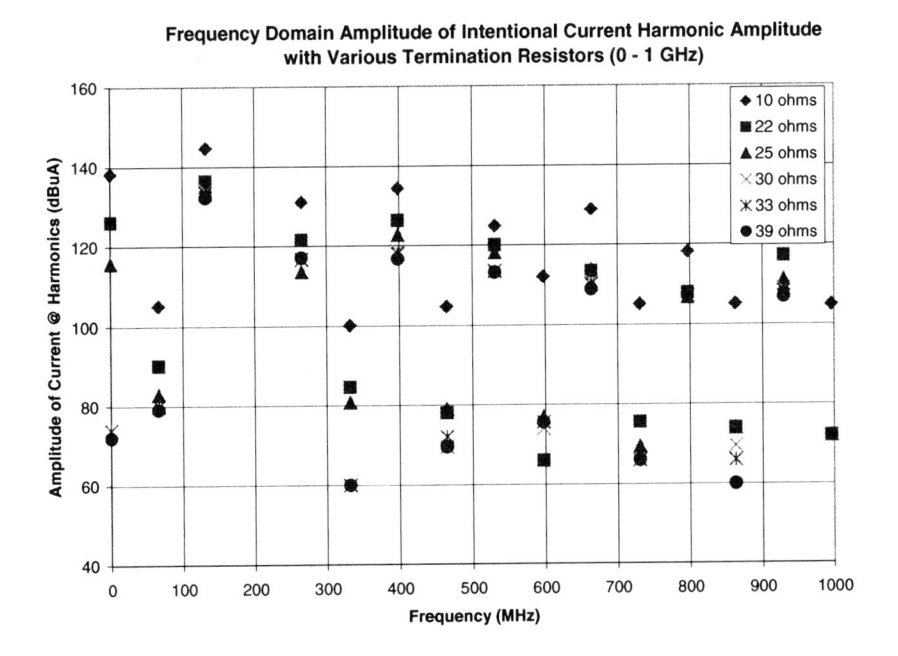

Figure 10-3 Harmonic Content of Current Waveforms (100-1000 MHz)

176 / PCB Design For Real-World EMI Control

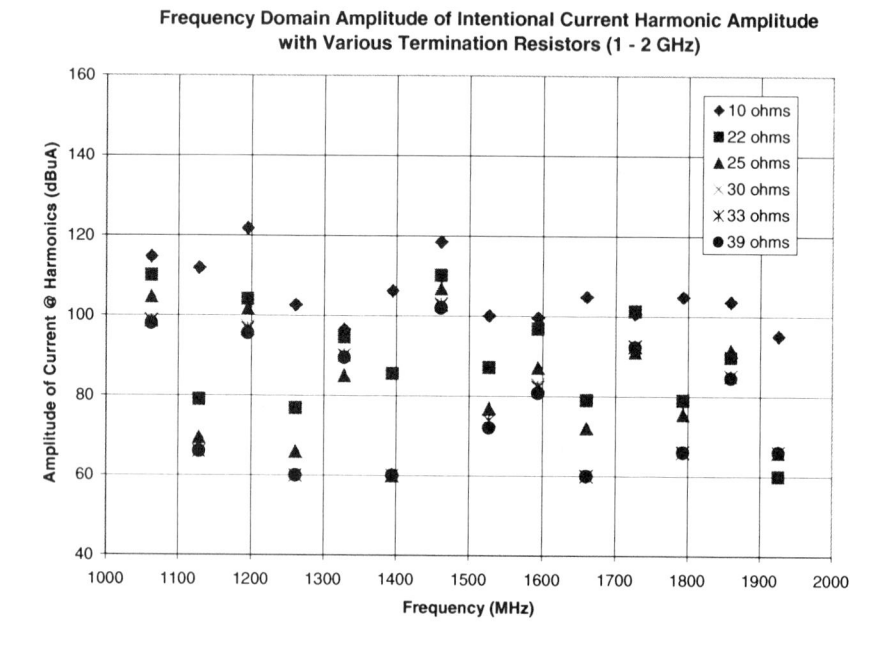

Figure 10-4 Harmonic Content of Current Waveforms (1000-2000 MHz)

Figure 10-5 shows the reduction in current amplitude (delta) for each harmonic frequency as the termination resistor value was varied. This figure also shows that the reduction in current amplitude is about the same for almost all harmonic frequencies, whether the termination resistor value is changed from 10 Ω to 39 Ω, or only from 10 Ω to 30 Ω. As can be seen in Figure 10-5, the amount of current reduction at some harmonic frequencies was as much as 45 dB! This is very significant, since few product designs fail EMI emissions standards by so great a margin. While a reduction in the harmonic current may or may not result in a one-for-one reduction in radiated emissions (depending on the exact coupling mechanism between the current and the final radiation source), there is still likely to be a significant reduction in emissions. This results in the need for much less filtering, gasketing, etc., of the final product. Engineers should learn to ask themselves, "Why fight an emission problem from a current that is not required in the first place?"

Frequency Domain Reduction in Intentional Current Harmonic Amplitude with Various Termination Resistors

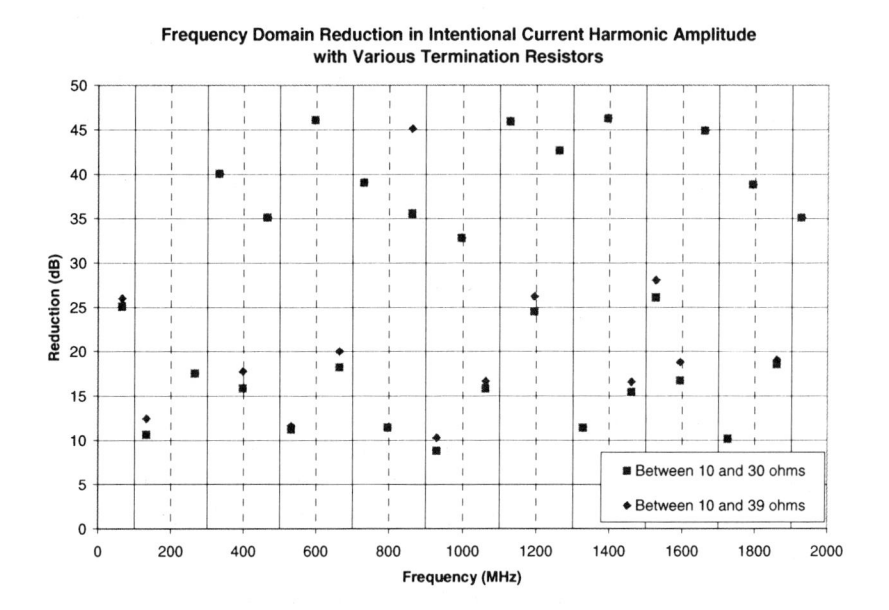

Figure 10-5 Reduction in Harmonic Amplitude of Current on Trace

10.3 Trace Current for Decoupling Analysis

When an IC/ASIC has a large number of output drivers, such as clock buffers, memory controllers, bus controllers, etc., the majority of current required to power the IC/ASIC is used to drive the I/O pins. Therefore, the current on the I/O traces is important, and can be used to analyze the decoupling capacitor design.

It was shown in [10.1] that the power pin current can be estimated by using the I/O current and the shoot-through current on clock buffers. A triangular pulse shape could be used for the power

178 / PCB Design For Real-World EMI Control

pin current from the I/O drivers. While this assumed waveshape is sufficient for some devices, sometimes the current waveshape is not as simple as the triangular pulse shape. The example in Figure 10-2 is evidence of this non-triangular waveshape.

Using signal integrity tools to find the currents on the I/O traces gives a good first order estimate of the current pulse at the power pin as well. In the case of clock buffers where the I/O pulses are synchronous, the current pulses can be directly summed to find the total current required through the power pins. In the case of a memory/bus controller, all of the I/O drivers are seldom driven at the same time, so using an average number of drivers operating at the same time provides a better estimate.

The type of termination technology used is also important. For example, typically SDRAM memory line uses a series source termination resistor only. The current from a typical SDRAM memory data line is shown in Figure 10-6. In contrast, a typical DDR RAM memory line uses a termination resistor connected to a voltage between the supply voltage and ground-reference. As shown in Figure 10-7, the current pulse shape driven from the DDR RAM I/O driver is quite different than that found for the SDRAM case.

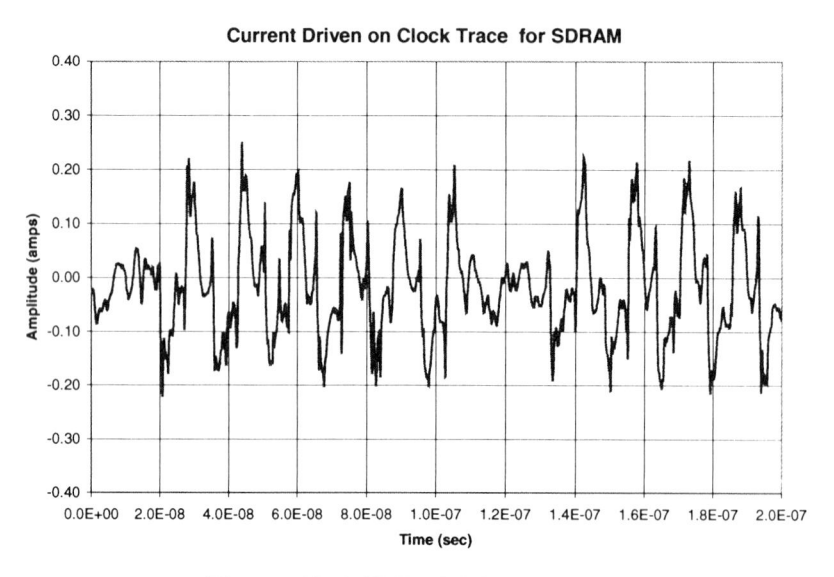

Figure 10-6 SDRAM Current Examp

Current Driven on Clock Trace for DDR RAM

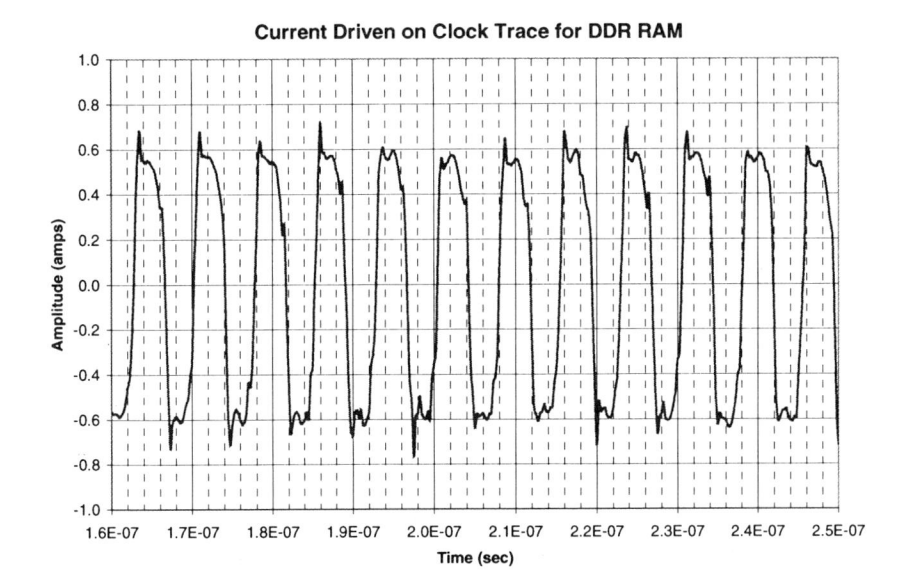

Figure 10-7 DDR RAM Current Example

Figures 10-8 and 10-9 show the frequency spectrum of the drive currents for SDRAM and DDRAM, respectively. Note that the harmonic amplitude is significantly different for the two types of memory. Simply looking at the voltage waveforms for the two types of memory would not have given any indication of the large differences in current harmonic spectrum.

10.4 Differential Signals Analysis

There are two types of differential signals that are important to EMC engineers: an internal high-speed line that is routed as a differential signal and a external I/O signal that is a differential signal. The EMC concerns are different, but can each be analyzed using commercial signal integrity tools.

180 / PCB Design For Real-World EMI Control

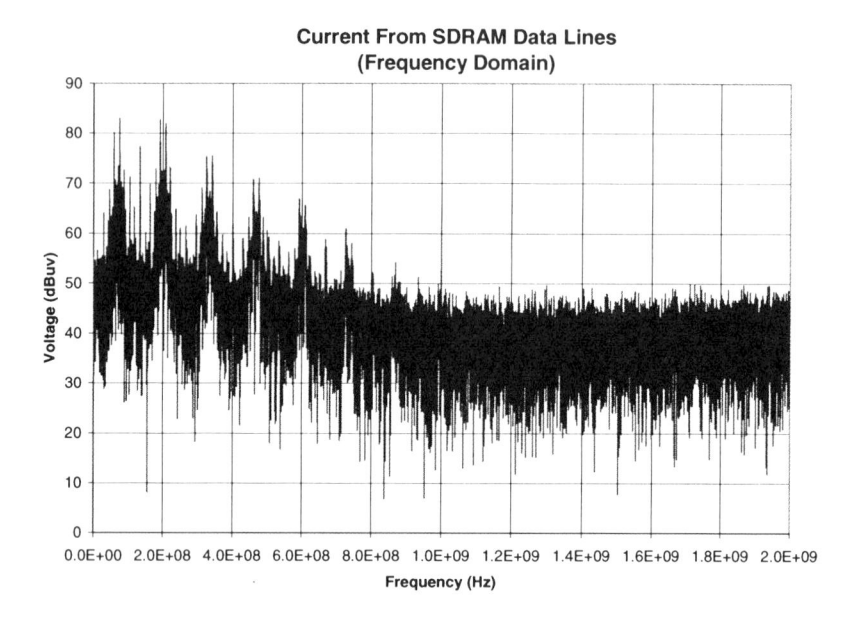

Figure 10-8 Example Frequency Domain for SDRAM Current

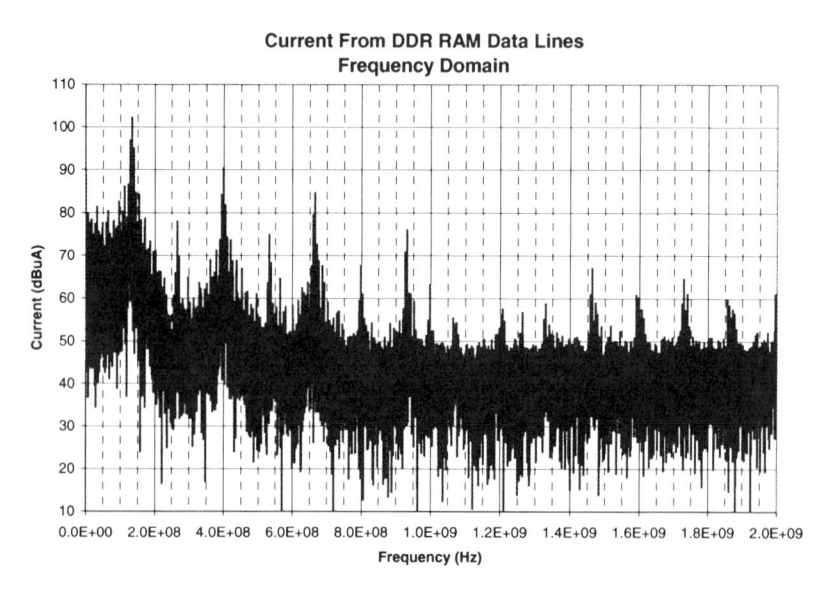

Figure 10-9 Example Frequency Domain for DDR RAM Current

10.4.1 Internal Differential Signal Lines

It is a common practice to use differential traces for high-speed signal applications on PC boards. These are not truly "differential", however, since they are closely coupled to the reference plane, and currents will flow in the reference plane. It is more accurate to consider these signals to be "complementary, single-ended" signals.[1] In Chapter 5, the need to keep single-ended signals from crossing splits in the reference plane, causing the return current to not be able to flow across the split, was discussed. It is often assumed that the currents in the reference plane from a differential signal pair will not be affected by the splits. If the length of the differential traces is identical there will be no common-mode currents on the traces. If the trace length is not identical (or the output drives are not properly balanced), then common-mode currents will exist on the traces. These common-mode currents must also return to their source, and they will use the reference plane. A split in the reference plane would not allow the return current for these signals to flow beneath the traces, and would cause the same problems as would a single-ended trace.

Signal integrity tools can be used to analyze the effect of mismatched length differential traces. For this example, the IBIS models for differential driver and receiver were used. The nominal transmission line length was set to ten inches, and then the length of one transmission line in the pair was increased slightly. The difference in the currents at the receiver is the common-mode current. Figure 10-10 shows the harmonic frequency spectrum of the common-mode current as the differential trace mismatch is increased. The amount of common-mode current with no mismatch is also shown as a base line.[2]

Figure 10-11 shows the increase on the common-mode current over the perfectly matched length case. Even for a minor mismatch of a few percent, some harmonics showed a significant increase in common-mode current. Since it is impractical to perfectly match the differential trace lengths on real world PC boards, and the driver is not likely to be perfectly matched as well, then differential traces

---

[1] Possibly called pseudo-differential signals.
[2] In this example, the baseline is non-zero because the output drivers were not perfectly balanced in the IBIS model file.

should be treated like single-ended traces and not run over splits in reference planes. An important point of this discussion is that the use of a commercial signal integrity tool can easily help analyze the effects of the trace length mismatch and provide valuable information to the design engineer.

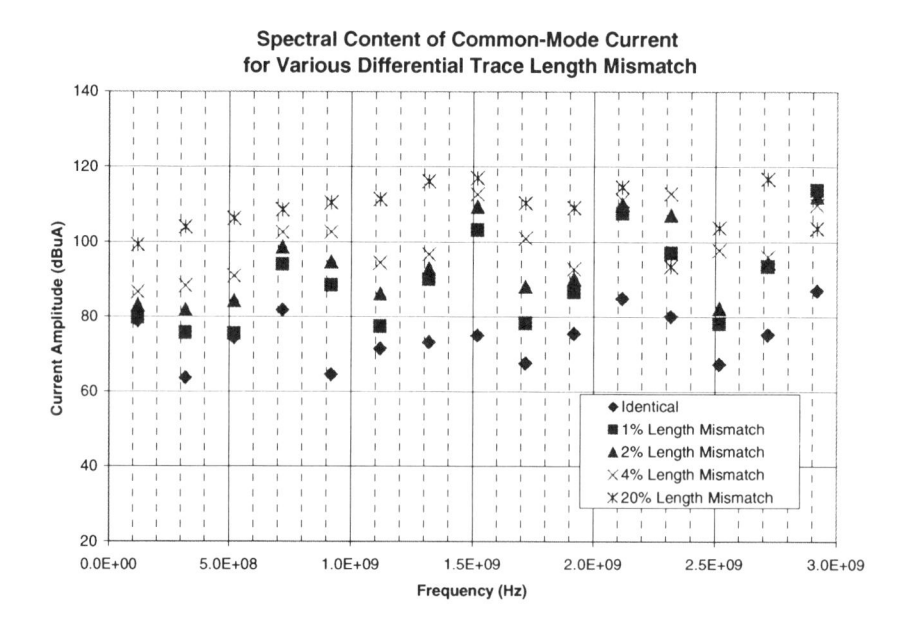

Figure 10-10 Common-Mode Current From Differential Traces

10.4.2 External I/O Differential Signal Lines

Differential signaling is common for certain types of I/O data, such as Ethernet and USB data signals. Commercial signal integrity tools can be useful to help analyze these data lines. For external I/O cables, the common-mode voltage between the cables and the enclosure chassis is a major concern for emissions control from the intentional signal. If the external cable provides the differential load to the traces, and the length of the traces is matched, then there will be no common-mode voltage between the cable and the chassis. As in the previous section, if the length of the traces is not matched, there will be a common-mode voltage on the cable as it exits the enclosure.

To analyze this common-mode voltage as the mismatch in the length of the traces increases, a circuit as shown in Figure 10-12 can be created. The differential drivers are differentially loaded by the cable impedance. In this case, the differential load is split into two halves, and the center of the two differential loads is connected to the ground-reference through a 100 ohm resistor. This value is typically a reasonable estimate for the common-mode impedance of an external I/O cable. When the length of the differential traces is matched, the common-mode voltage across the 100 ohm resistor will be zero. As the length mismatch increases, the common-mode voltage will increase also. Figure 10-13 shows an example of the increase in the harmonic content of the common-mode voltage as the trace length mismatch is increased.

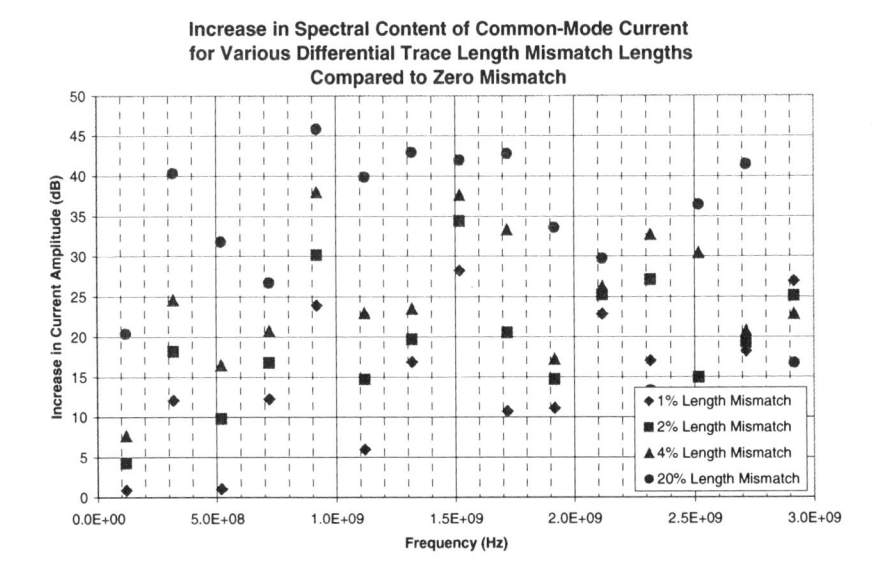

Figure 10-11 Increase in Common-Mode Current From Differential Traces as Trace Length Mismatch Increases

184 / PCB Design For Real-World EMI Control

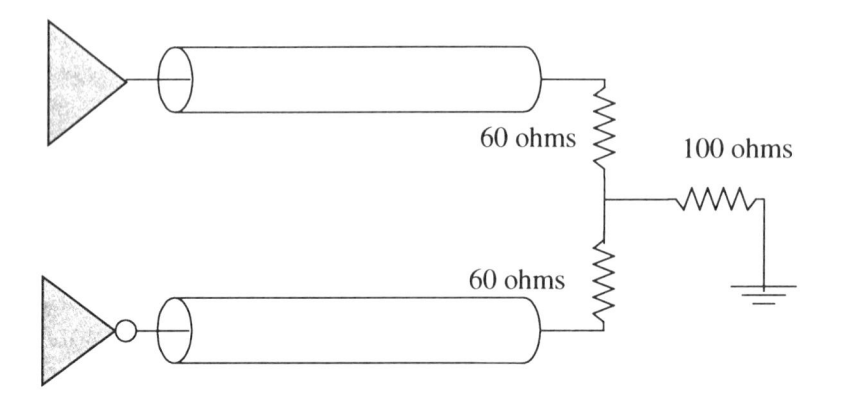

Figure 10-12 Equivalent Circuit for Common-mode Current from Differential Traces

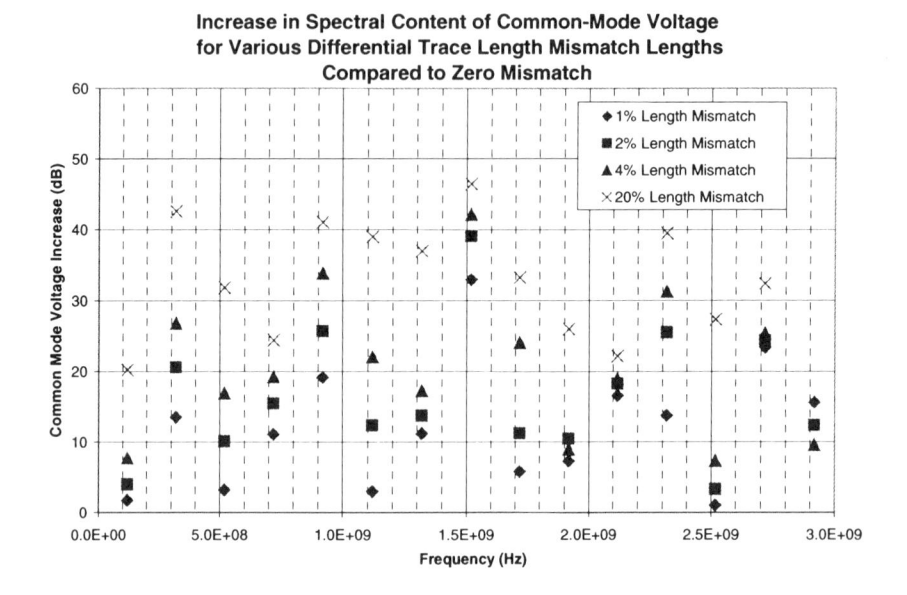

Figure 10-13 Increase in Common-Mode Voltage From Differential Traces as Trace Length Mismatch Increases

10.5 Crosstalk Analysis

Signal integrity tools are commonly used for crosstalk analysis. Normally this analysis is only used to determine if the level of crosstalk noise that occurs will interfere with proper data transfer. This analysis is most often done only between high-speed signal lines, within high-speed buses, etc. [10.2] Crosstalk between high-speed signal lines and I/O lines is seldom considered, unfortunately, yet this form of crosstalk concerns EMC engineers most.

Signal integrity tools allow users to include filter components in their circuit models, so analysis of the high-speed signal trace to the I/O trace can be accomplished and the effects of filters included. In this manner, the level of common-mode voltage on the I/O connector can be determined for a number of different internal sources.

10.6 Summary

There are a number of useful analyses that can be accomplished by using commercial signal integrity tools for EMC emissions control. Most designs already undergo some degree of analysis using signal integrity tools, so it is straightforward to extend this analysis for certain EMC applications.

The most important analysis that should be performed is to ensure that the frequency spectrum of the intentional current on a trace is as low as possible while still meeting the data transfer requirements. Voltage waveform analysis alone is not sufficient to determine the current waveform. Some commercial signal integrity tools allow analysis of the current. Investigating the behavior of this current allows the termination to be optimized to keep the high-frequency harmonics low, thus eliminating a possible problem at its source.

Other standard signal integrity tool analyses can also be applied for EMC. Crosstalk analysis between high-speed traces and I/O traces, analysis of common-mode current from mismatched differential traces, and analysis of common-mode voltage on differential I/O lines can all be easily performed.

References

[10.1] J. Mao, S. Luan, B. Archambeault, J. Fan, J. L. Drewniak, and T. P. Van Doren, "Estimating DC power bus noise in multi-layer PCBs using IC transient current and the power plane impedance, *IEEE EMC Symposium Proceedings*, Minneapolis, MN, August 2002.

[10.2] W. Cui, M. Li, X. Luo, J. L. Drewniak, T. H. Hubing, T. P. Van Doren, R. E. DuBroff, "Anticipating EMI from coupling between high-speed digital and I/O lines," *IEEE Electromagnetic Compatibility Symposium Proceedings*, Seattle, WA, pp. 189-194, August 1999.

Chapter 11

Printed Circuit Board Layout

11.1 Introduction

The initial layout of a printed circuit (PC) board is an important part of the overall EMC strategy. While for some aspects of the design there is very little that can be negotiated, it is worth the time and effort to consider what can be done.

Mechanical constraints may require certain connectors to be positioned along one side of the board. Processors with heatsinks may be required to be next to an air vent, placing the processor near the edge of the board. Electrical signal integrity constraints may require certain IC/ASICs to be close to one another. Cost constraints may require the PC board to only include two power/ground-reference planes.

All systems have some constraints, but there is usually some amount of allowable variations, and unless EMC is considered, it is likely that a non-optimum decision will be made. Many of these issues are covered in various other chapters in this book. However, these issues are important enough to bring them all together in one chapter.

11.2 PC Board Stack-up

The stack up of the PC board is most often determined by the target cost of the board, the manufacturing technology, and the number of wiring channels required for functionality. As with most engineering designs, there are a number of conflicting requirements, and the final design strategy is decided after considering a number of trade-offs. PC boards with anywhere from 1 layer for the simplest, lowest cost boards, to 30 or more layers for high-performance systems, can be used.

11.2.1 Many Layer Boards

PC boards with many layers are most often used for high-speed, high performance systems. A number of layers are used as either DC power or ground-reference planes. The planes are most often solid planes with no splits, since there are usually enough different plane layers that there is no need to have different DC voltages on a single layer. Regardless of the schematic 'name' for a layer (e.g. "ground", +5 volts, VCC, digital power, etc), the plane will act as the return current path for the signals in the transmission lines directly adjacent to them. Creating a good low impedance return current path is the most important EMC task for these plane layers.

Signal layers are interspersed between the solid plane layers. They can be either strip line, where the spacing between the signal layer and the two adjacent plane layers is equal, or they can be asymmetrical striplines where there are two signal lines sandwiched between adjacent plane layers. Figure 11-1 shows an example of the cross section for both cases. In most designs, a number of different combinations of these configurations are used.

Stripline Configuration   Asymmetrical Stripline Configuration

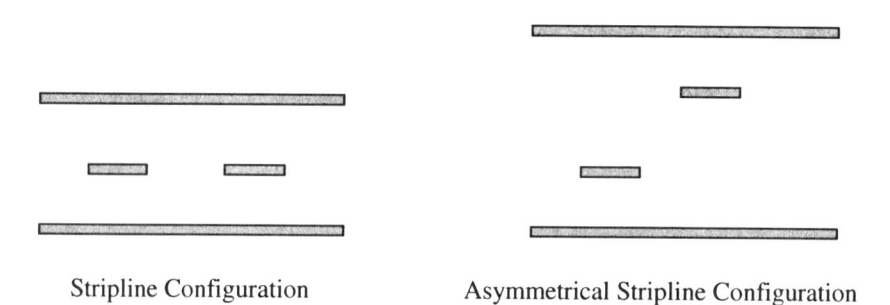

Figure 11-1 Symmetric and Asymmetric Stripline Configurations

One popular PC board stack-up for a 12 layer board is T-P-S-P-S-P-S-P-S-S-P-B, where "T" is the top layer, "P" is a plane layer, "S" is a signal layer, and "B" is the bottom layer. The top and bottom layers are used for component pads, and signals should not be run on them for any significant distance in order to reduce direct emissions from the traces. While this example is only one stack-up, the design considerations and thought processes can be extended to any other stack up configuration.

The next consideration is to determine which, if any, plane layers will have to contain multiple power islands for different DC voltages. For this example, assume that layer #11 (from the top) will have multiple DC voltages. This means that designers must keep the high-speed signals away from layers #10 and the bottom as much as possible, since the return current cannot flow across the gap in the planes on layer #10 and stitching capacitors would be required. This leaves layers #3, #5, #7, and #9 as signal layers for high-speed signals.

The next consideration is to plan the routing of the most critical signals. In most designs, the traces are laid out in one direction as much as possible to optimize the number of available wiring channels on the layer. Layers #3 and #7 might be assigned as "east-west" and layers #5 and #9 assigned as "north-south". Which layer a trace is routed on depends upon the direction it needs to travel to reach its destination.

An important consideration is the changing of layers for a high-speed trace, and which different layers will be used for an individual trace. Again, the major consideration is to insure the return current can travel from one reference plane to the new reference plane where it is intended to flow. In fact, the best design will not require the return current to change reference planes, but to simply change from one side of the plane to the other side. For example, the following combinations of signal layers can be used together as signal layer pairs: #3 and #5, #5 and #7, and # 7 and #9. This allows both an east-west and a north-south routing in each combination. A combination like #3 and #9 should not be used because this would require the return current to flow from the plane on layer #4 to the plane on layer #8. While a decoupling capacitor could be placed near the via, the capacitor will not be effective at high frequencies due to lead and via inductance. Such wiring, by forcing the need for a capacitor, also increases the part count and the cost of the product.

Another important consideration is to assign the DC voltages for the plane layers. Let's suppose for this example that the processor is expected to have the most significant amount of noise on the power/ground-reference pins because of the high-speed nature of the internal processing. This means that it is very important for the decoupling capacitors assigned to the same DC voltage as the processor to be as effective as possible. As shown in the chapter on decoupling, the high-frequency performance of the on-board decoupling capacitors was severely limited by the inductance of the connecting vias, pads and connection traces. The best way to lower this inductance is to keep the connecting traces short and wide, as well as keeping the vias short and fat. If layer #2 is assigned as "ground" and Layer #4 is assigned as processor-power, then the via distance from the top where the processor and the decoupling capacitors are installed is as short as possible. The remaining part of the via that extends down to the bottom of the board does not include any significant currents, and is too short to be considered a 'stub' antenna. Figure 11-2 shows an illustration of this stack-up design, and it can be seen that placing a capacitor on the bottom of the board would have resulted in longer vias and thus greater inductance than placing the capacitors on the top layer.

The next important consideration is *which* high-speed signal lines will be routed on layers #3 and #5. It is desirable to keep the signal traces that are driven from an active device with the same power as the reference plane together. That is, signals from the processor (for example, the memory bus and other high-speed buses) should be routed on layers #3 and #5 since they share the same power and the return currents can thus more easily return to their source.

While this section has focused on only the most critical signals and IC (for this example), the above thought process should be continued for the remaining signals and ICs. The signals routed on layer #10 should only be low speed signals because of the split in the planes on layer #11.

Printed Circuit Board Layout / 191

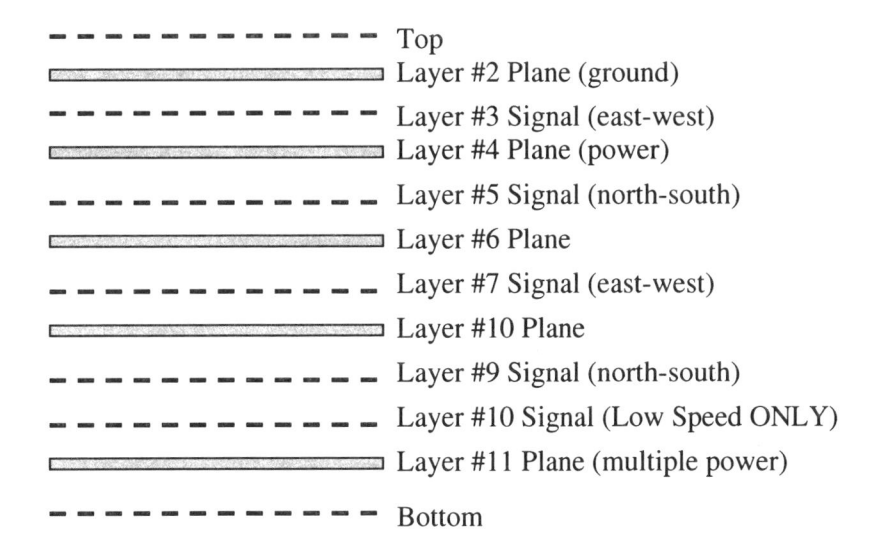

```
― ― ― ― ― ― ― ― ― ― ― ―  Top
▭▭▭▭▭▭▭▭▭▭▭▭  Layer #2 Plane (ground)
― ― ― ― ― ― ― ― ― ― ― ―  Layer #3 Signal (east-west)
▭▭▭▭▭▭▭▭▭▭▭▭  Layer #4 Plane (power)
― ― ― ― ― ― ― ― ― ― ― ―  Layer #5 Signal (north-south)
▭▭▭▭▭▭▭▭▭▭▭▭  Layer #6 Plane
― ― ― ― ― ― ― ― ― ― ― ―  Layer #7 Signal (east-west)
▭▭▭▭▭▭▭▭▭▭▭▭  Layer #10 Plane
― ― ― ― ― ― ― ― ― ― ― ―  Layer #9 Signal (north-south)
― ― ― ― ― ― ― ― ― ― ― ―  Layer #10 Signal (Low Speed ONLY)
▭▭▭▭▭▭▭▭▭▭▭▭  Layer #11 Plane (multiple power)
― ― ― ― ― ― ― ― ― ― ― ―  Bottom
```

Figure 11-2 T-P-S-P-S-P-S-P-S-S-P-B Stack-up

11.2.2 Six-Layer Boards

A common special case is the six-layer board. Typically, this stack-up is selected for lower cost products and contains 4 signal layers and 2 plane layers. Figure 11-3 shows an illustration of this stack-up. There are typically two plane layers and four signal layers. Obviously this provides a lot less freedom than in the previous case, but designers can still make some selections that will help the EMC performance of the system.

As in the previous case, an east-west and north-south routing is normally used. Again, it is desired to use routing layer pairs that do not require the return current to change planes. In this case, we'll select layers #1 & #3 as a routing pair and layers #4 & #6 as the other routing pair. In this case, the top and bottom layers must be used for routing signals. For good EMC performance, it is more important to keep the return current on a single plane than to keep the signal buried between planes, therefore, layers #3 & #4 should *never* be used as a routing pair for high-speed signals.

Layer #2 and layer #5 will be the power and ground-reference layers. It is very likely that there will be a number of different DC power requirements, so the power plane is most likely to be split up into a number of power islands. If layer #2 is selected as the ground-reference layer, then designers must insure that all the high-speed signals are routed on layers #1 & #3 so that they do not cross splits in the reference plane. Naturally, if a particular signal's path does not take it across a split in the (power) reference plane, then routing these signals on layers #4 & #6 is acceptable.

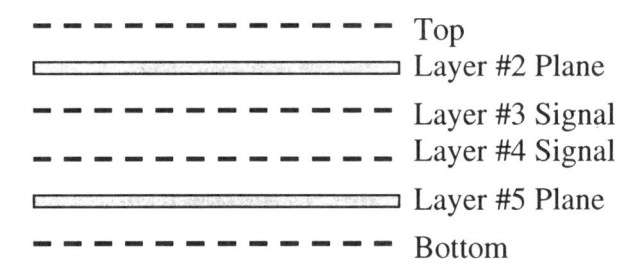

```
- - - - - - - - - - - -   Top
[=========================]  Layer #2 Plane
- - - - - - - - - - - -   Layer #3 Signal
- - - - - - - - - - - -   Layer #4 Signal
[=========================]  Layer #5 Plane
- - - - - - - - - - - -   Bottom
```

Figure 11-3 T-P-S-S-P-B Stack-up

11.2.3 Four-Layer Boards

Four-layer boards are used for low cost systems whenever possible. Typically, there are only two signal layers and two plane layers. Figure 11-4 shows an illustration of this stack-up.

Optimization of the number of routing channels is of critical importance, so the east-west, north-south routing strategy is used. This time, however, it is not possible to maintain the same reference plane for the return current. A decoupling capacitor must be placed close to the via to provide a return current path. The short trace connecting between the capacitor pad and the via should be maintained as short as possible and as wide as possible to keep the inductance/impedance to a minimum.

The planes are normally assigned as ground-reference and power, with the power plane again being split into a number of different voltages. It is extremely important to maintain the traces over solid plane areas so that traces do not cross these splits when the power plane is used as the signal reference for a trace. If split crossing does occur, a stitching capacitor must be placed closely to the point where the trace crosses the split to provide a path for the return current. Again, the short trace connecting between the capacitor pad and the via should be maintained as short as possible and as wide as possible to keep the inductance/impedance to a minimum.

```
– – – – – – – – – – – Top
▭▬▬▬▬▬▬▬▬▬▬▭ Layer #2 Plane


▭▬▬▬▬▬▬▬▬▬▬▭ Layer #3 Plane
– – – – – – – – – – – Bottom
```

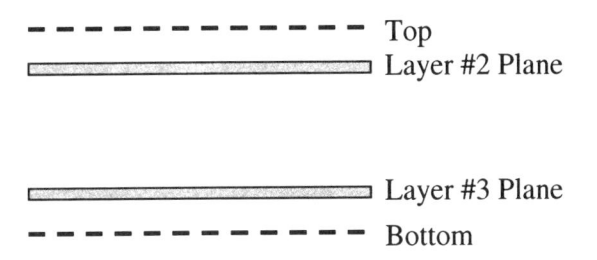

Figure 11-4 T-P-P-B Stack-up

11.2.4 One and Two-Layer Boards

One and two-layer boards present a definite challenge for EMC design. While this stack-up is not recommended, it is often selected for considerations other than EMC. With this stack-up, there are normally no solid planes and all signals and all power and power return are routed as traces. The main concern in this design strategy is to keep the loop area for the signal current as small as possible and not allow large loop areas for current as illustrated in Figure 11-5.

While the signal speeds on these boards are normally less than might be expected on the multilayer boards in previous sections, they can still cause EMC problems. A signal return trace should be routed along side the signal trace to minimize the loop area and therefore minimize the emissions (and susceptibility to external RF interference). Figure 11-6 shows an illustration of this design strategy. Keeping signal traces short will also minimize the loop area.

Decoupling capacitors should be positioned as close to the ICs as possible and connected between the power and ground pins.

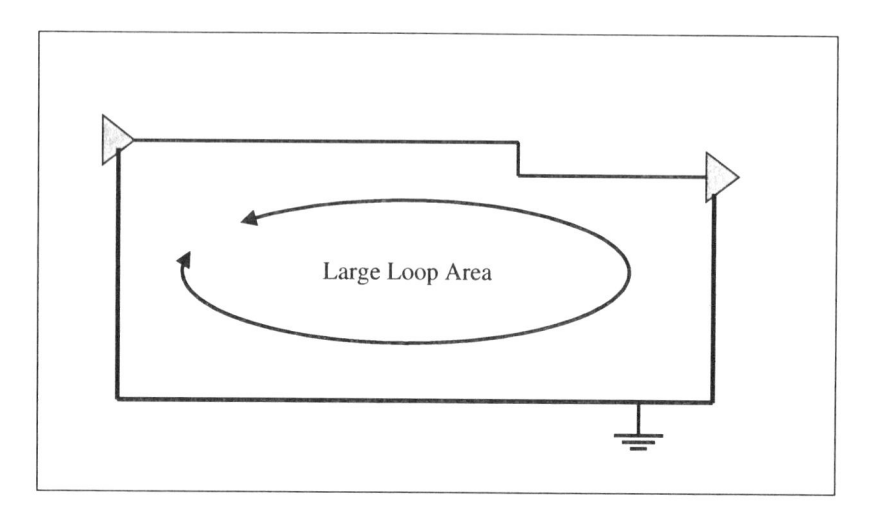

Figure 11-5 Example of Single Sided PCB Routing with Large Loop Area for Signal Current

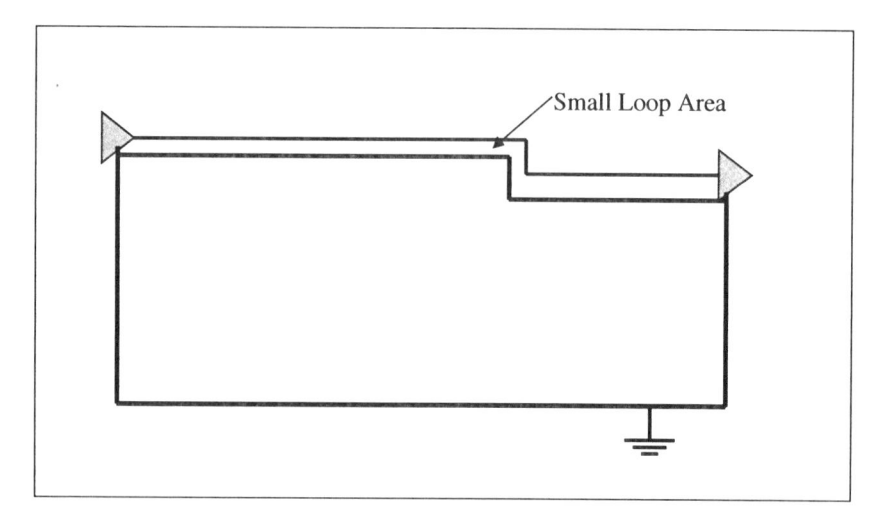

Figure 11-6 Example of Single Sided PCB Routing with Small Loop Area for Signal Current

11.3 Component Placement

As mentioned earlier, the location of many of the components is often decided by other factors. Whenever possible, there are a few considerations that can help the EMC performance of the system.

An important rule is to keep high-speed devices together in one region of the board and to keep low speed devices together in another region, and to keep the high and low speed devices separated as much as possible. It is possibly even more important to consider where the signals from these devices will be routed on the board. For example, the I/O connectors are usually fixed along one side of the board. It is important to keep the I/O driver devices close to the I/O connectors so that unwanted noise from high-speed devices cannot couple onto the I/O traces and be conducted out of the enclosure. If the I/O driver is placed far from the connector area, then the likelihood of this coupling is greatly increased.

High-speed devices will usually drive high-frequency currents on the device pins. These currents create electric and magnetic fields which can couple directly onto I/O connector pins when the high-speed device and the I/O connector are placed close together. None of the filters on the board can stop signals coupled directly onto the I/O connector pins because the coupling occurs past the filter location.

There are a number of ways for undesired coupling to occur. It is not possible to consider every possible combination of devices traces, etc., here. A little design consideration can help reduce costs and problems, however, and should be encouraged for whatever the design.

11.4 Isolation

The previous section explained why high-speed circuits should be kept away from low speed and I/O circuits. Unfortunately, this is not always possible because of other design requirements. Some amount of additional isolation is needed even if the different circuits are fairly close together.

One of the greatest concerns is the spread of return current in the reference planes resulting from the routing of high-speed traces and placement of high-speed devices. This return current in the reference plane can be isolated from low speed circuit areas by using intentional splits in the reference plane. Naturally, this is not a desirable strategy if the reference plane is adjacent to a layer where intentional high-speed signals are routed which would then cross this split (as described in Chapters 5, 6, and 7). If there are no traces routed over the area where an isolation split would be used, this can be very effective at blocking unwanted noise from reaching low speed and I/O devices up to several hundred megahertz.

One example of intentional split use is that which isolates the ground-reference plane in the I/O connector area from the digital ground-reference. This is explained in more detail in Section 7.4. This design strategy keeps the high-frequency noise caused by the high-speed digital circuits from spreading into the I/O area, but keep in mind that a low frequency return current path must be provided across the split in the reference plane for the intentional I/O currents.

Isolation of particular devices is possible by creating a 'moat' in the power plane and the ground-reference planes around a particular device. Again, extreme care is needed so that high-speed signal traces are not routed across this moat and sufficient path for the DC power current must also be provided. The location of the splits creating this "moat" on the power and ground-reference planes must be identical in order to be effective and prevent accidental coupling between them.

11.5 Summary

The number of layers, number of solid planes, and configuration of the signal layers on a PC board stack up is most often decided without thought to EMC concerns. By judiciously selecting which layers to route which signals, the return currents can be allowed to remain near the appropriate signal traces and the EMC performance greatly improved over that created by random routing. This careful routing does not increase board cost, and while it may take a slightly longer time to route the board, this is more than offset by the reduction in EMC problems. Consideration of the return current path is extremely important and is probably one of the most over-looked concepts in good EMC design.

Chapter 12

Shielding in Enclosures with Apertures

12.1 Introduction

The fast data and edge rates of modern high-speed digital electronics often necessitate the use of some type of shielding to comply with EMI regulatory or other design requirements. The specifics of the shielding for a particular type of product and the expectations for EMI reduction can vary widely. At one end of the spectrum might be the transmitter or receiver of a wireless communication link operating in the microwave region. In this case, the shielding enclosure will typically be a cast or machined conductive enclosure. The distinct pieces of the enclosure might be screwed together, with careful attention devoted to the seam geometry at the microwave design frequency in order to prevent leakage from the interior or interference from external sources. All penetrations of the shielding enclosure will be carefully designed. Signals into or out of the design will be on shielded cables, with connectors that have 360^0 connections, e.g., a coaxial connector. Any microwave signals will be on coaxial cables. An electronic design requiring this level of attention to the shielding enclosure will be one for which cost is not an issue, or the design cannot function without such intricate attention to every enclosure penetration.

At the other end of the design spectrum are those products that are driven entirely by cost, and using any shielding enclosure, much less the expense associated with the enclosure described above, would be cost-prohibitive. In these cases, considerable effort is devoted to careful EMC design at the PCB level, and possibly only a few decibels of attenuation is provided by the shielding needed to meet EMI compliance and margin requirements. A sheet-metal plate closely spaced to the component side of the PCB may provide a few decibels of attenuation to meet requirements, or a local shield over a specific component may help achieve compliance. A myriad of design possibilities lies between these two shielding extremes and might include plastic enclosures sprayed with a conductive paint, carbon-fiber-composite enclosure materials, shielded cables, and metal connector shells, among many other considerations. The design requirements will be driven predominantly by the EMI attenuation budgeted for the shielding enclosure and cost. Expectations for high-frequency shielding performance at megahertz and gigahertz frequencies range from 5-30 dB.

A typical shielding enclosure for a high-speed digital product is illustrated in Figure 12-1. A cost-effective shielding enclosure constructed of sheet metal will often have numerous perforations and penetrations required for other design aspects that compromise the shielding effectiveness. The slots, seams, and apertures that result from construction of the shielding enclosure, plug-in modules, and airflow will be a dominant source of leakage. In addition, the cable penetrations and radiation from common-mode currents on the cable will also be a significant contributor to the overall emissions. Typical design guidelines based on "best practice" experience mandate slot lengths less than one-tenth of a wavelength at the highest frequency harmonic that is anticipated coming from the interior electronics, shielded cables, and/or filtering of unshielded cables. These design guidelines are inadequate for enclosures of leading-edge electronics. To minimize the electrical dimensions of perforations, screws, finger-stock, or conductive gaskets are often used to provide electrical continuity across a slot, seam, or joint in the metal enclosure. In practical applications, the limited effectiveness of these contacts then dictates the true shielding effectiveness of the enclosure. Similarly, the performance of shielded cables is to a large degree determined by the connector-enclosure interface. In particular, the cable shield must be connected to the metal connector shell, and this shell must in turn be connected to the conducting face of the shielding enclosure. The EMI performance of the shielded cable is typically dominated by this interface, and less often the cable shield braid density.

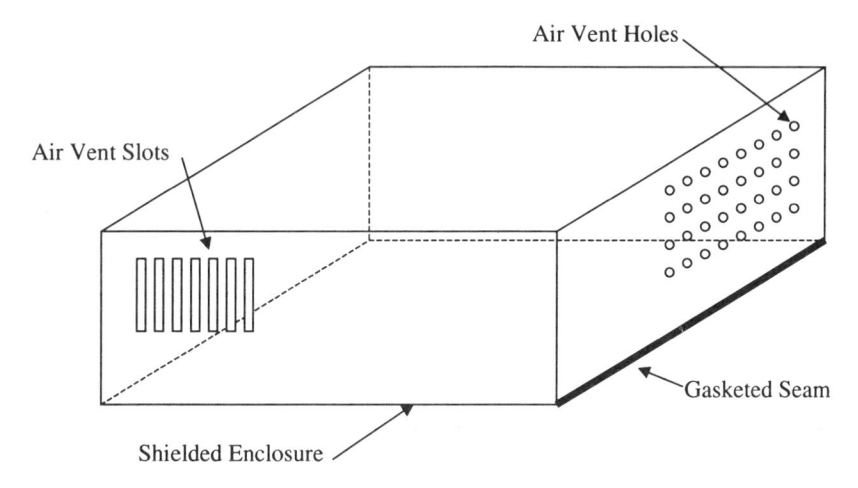

Figure 12-1 A Typical Shielded Enclosure

The ideal shielding enclosure would be a metal case, with a metal or sprayed conductor thickness that is several skin depths thick at the highest frequency at which attenuation is needed. With this perfect shielding enclosure, all parasitic currents, i.e., common-mode currents, which wind up on the enclosure walls are confined to the interior surface of the enclosure. There are no paths through gaps in the enclosure for conduction currents on the interior wall to gain access to the outer enclosure wall surface. Intended currents that enter or leave the enclosure do so on a shielded cable that has a perfect 360^0 connection at the connector-enclosure interface, and the connection on the other end of the cable is similarly perfect. In this fashion, the parasitic common-mode currents that are inevitable in a real design are confined to the interior of the perfect shielding enclosure and do not lead to radiated EMI. The reality of practical, cost-effective shielding enclosures is quite different. The EMI performance of the shielding enclosure will be dictated by the perforations resulting from gaps, seams, and airflow, as well as the cable penetrations and the resulting common-mode currents on cables and the consequent EMI. In addition, the performance of gasketing materials, and the ability to establish metal-to-metal contacts between the gaskets and the shielding enclosure conducting walls will impact the EMI performance. The frequency dependence of absorbing materials in wire-knit gaskets or absorbing materials used elsewhere in the design will also have an effect. Finally, the particular structures on the PCB in the enclosure interior that comprise the EMI coupling path and have the common-mode currents on them will impact the EMI performance of a specific shielding enclosure design. These aspects of practical shielding enclosure design are addressed in this chapter.

12.2 Resonance Mode within Shielded Enclosures

The radiated EMI resulting from an imperfect shielding enclosure design is caused either by radiation through an aperture or radiation due to common-mode currents on cables. A conduction current, i.e., current carried in the conductor by electrons, encountering a slot discontinuity on a conductor surface is illustrated in Figure 12-2. The two independent cases to be considered are for conduction currents that are perpendicular to the slot, denoted the TE (transverse electric) polarization, and for conduction currents that are parallel to the slot, denoted the TM (transverse magnetic) polarization. In the TE case, the conduction current flow encounters the slot perpendicular to the slot orientation. The conduction current must part and flow around the slot as depicted. Displacement current, i.e., current carried by a time-changing electric field, results across the slot. It is the displacement current, or the electric field perpendicular to the slot that leads to significant radiation when the dimensions of the slot or the aperture become appreciable relative to a wavelength. The radiated electric field has the same polarization as the electric field across the slot. In the TM case, the conduction current flows along the slot axis, and the resulting perturbation in the current path is small. Consequently, the radiation resulting from the TM case is several orders of magnitude less than that for the TE case and is not of concern. This was also the principle behind slotted waveguide or slotted coaxial cable line measurements for standing wave ratio prior to modern instruments. The slot in the waveguide was cut along the axis of the current flow, and so perturbed the conduction current, and hence, the fields that were being measured negligibly [12.1].

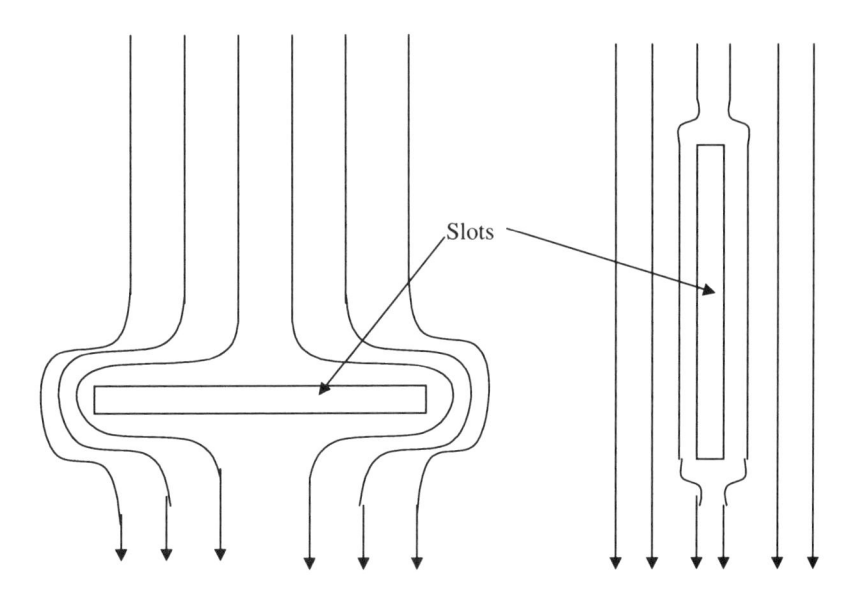

Figure 12-2 Conduction Current Around Slots in Metal Shield

A conducting enclosure supports resonant modes at which the electromagnetic fields can potentially become large. If these large fields excite a slot in the enclosure wall, appreciable EMI can result [12.2], [12.3], [12.4]. A resonant mode within an enclosure is analogous to the resonance frequency of an RLC circuit. For example, consider the parallel RLC circuit shown in Figure 12-3.

At the parallel resonance frequency of the circuit,

$$\omega_o = \frac{1}{\sqrt{LC}} \qquad (12.1)$$

the voltage at the circuit terminals that results from a sinusoidal current source applied at this frequency becomes very large as shown in Figure 12-4. In fact, the response is only limited by the loss in the circuit from the resistive element R. This is usually referred to as the "natural response" of the circuit [12.5]. Analogously, a conducting enclosure will have resonant frequencies at which the input impedance seen at the terminals of a source driving the enclosure will look like that of Figure 12.4. A source driven at a resonant frequency will produce a large response, or electromagnetic fields, within the shielding enclosure that in turn can lead to EMI if this energy is coupled to the exterior through a perforation or onto a cable.

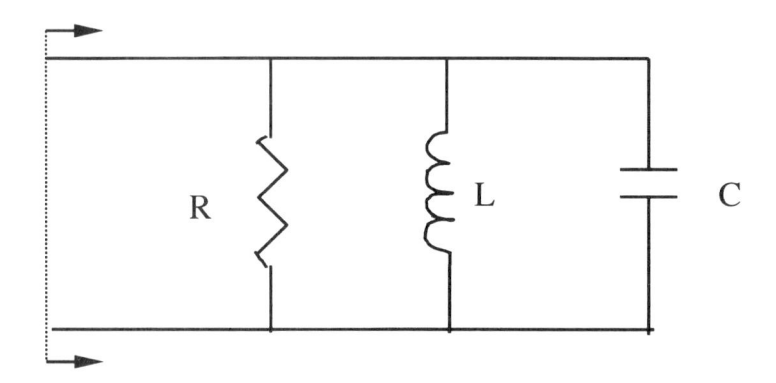

Figure 12-3 Parallel RLC Circuit Schematic

The modes for a conducting enclosure are a function of the enclosure geometry and internal medium. The electric and magnetic fields as a function of space and frequency, as well as the resonant frequencies, can be calculated analytically for special cases that have enclosure walls that conform to a rectangular, cylindrical, or spherical geometries (in addition to some other peculiar shapes for separable geometries) [12.6]. The simplest case to consider is that of a rectangular geometry. The modal electric and magnetic fields are calculated from the source-free Maxwell equations, and basically correspond to fitting integer numbers of half-wavelengths along each spatial dimension. The solutions to Maxwell's equations are waves, and in the case of a resonant enclosure, take the form of standing waves and are sine and cosine functions. A half-wavelength is one half of a period in space of a trigonometric function. This basic concept is illustrated simply in Figure 12-5 for the E_y component of the TE_{101} and TE_{102} modes within a rectangular shielding enclosure.

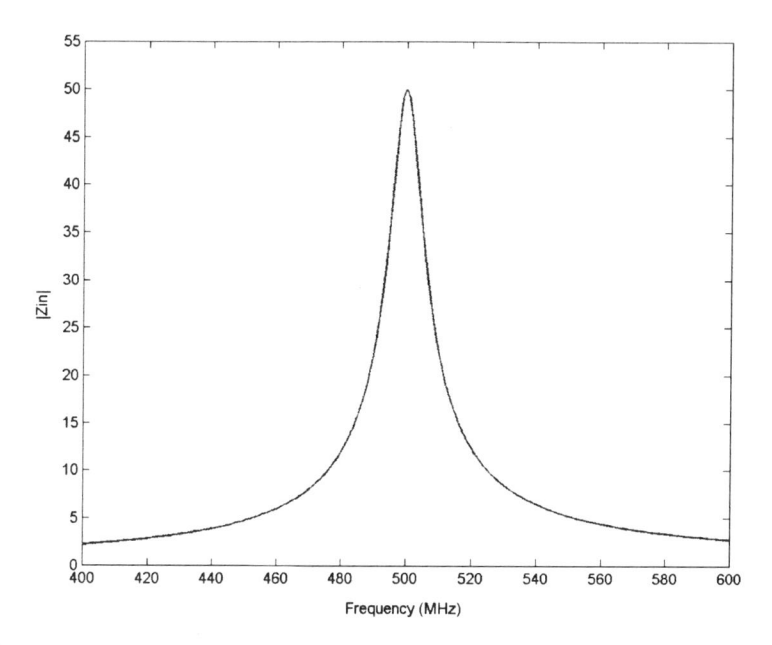

**Figure 12-4 Parallel RLC Circuit Input Impedance for Q=50
R=50 Ω, C=318 pF, L=0.32 nH**

Here the designation TE corresponds to the electric field being entirely transverse to the z-axis, and the subscripts (1,0,1) designate the number of half-wavelength sinusoidal variations there are in the fields along the (x, y, z) axes. The particular sinusoidal function, sine or cosine, is chosen so that the boundary conditions on the electric field are met. Referring to Figure 12-5, the boundary conditions dictate that the component of the electric field tangent to a conductor must go to zero on the surface of the conductor [12.7], thus E_y must be zero along the $x=0,a$, and $z=0,d$ faces of the enclosure. In general, the spatial variation of the modal fields within the enclosure can get complicated, but with the above simple notion, the resonant frequencies of the modes are easily determined as [12.7]

$$f_{mnp} = \frac{1}{2\sqrt{\varepsilon\mu}} \sqrt{\left(\frac{m}{a}\right)^2 + \left(\frac{n}{b}\right)^2 + \left(\frac{p}{c}\right)^2} \qquad (12.2)$$

where:
a, b, and c = length of the sides of the cavity, and
m, n, and p = integer numbers (only one can be zero at a time),

and are only a function of the geometry and medium filling the enclosure.

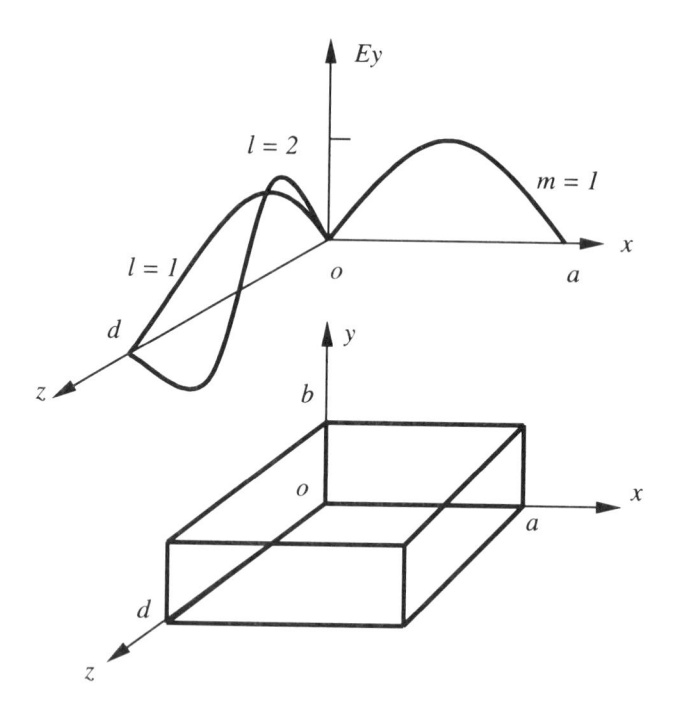

Figure 12-5 Low Order TE_{101} and TE_{102} Modes Within a Rectangular Shielded Enclosure

A particular mode within an enclosure is excited by electric and magnetic sources that are located near the peaks of the mode, and will not be excited by a source placed near a null in the modal field. For example, an electric current source oriented along the y-axis, and placed at $z=(1/2)d$ will excite the TE_{101} mode within the enclosure, but not the TE_{102} mode. The TE_{101} mode will be excited because the electric current source at $z=(1/2)d$ is at a maximum of the electric field for this mode, but the TE_{102} will not be excited by this electric source, because the source is placed at a null of the electric field for this mode. Both electric and magnetic sources can excite modes within the cavity, however the electric sources excite the electric field and the magnetic sources excite the magnetic field. An electric source is simply an electric dipole, and a magnetic source is a current loop. To excite a particular mode, the electric sources are placed at or near a maximum of the electric field, and the magnetic sources are placed at or near a maximum in the magnetic field. In fact, rigorously, it is the inner product of the source (polarization and spatial function) with the modal fields that dictates how strongly, or if, a particular mode will be excited [12.6].

From a practical point of view, products are rarely empty rectangular cavities. A variety of metal boxes for power supplies, disk drives, and other subsystems are positioned around the interior of the enclosure. A number of PC boards are positioned in different orientations and wires and cables are used to interconnect the various boards and subsystems. The boundary condition whereby the tangential electric field must be equal to zero on perfect conductors described earlier must still be met, and this creates modes that are much more difficult to visualize.

Consider the example of an enclosure with the dimensions such that the first resonant mode occurs at 400 MHz. As described earlier, the electric field mode that supports this resonant frequency will have zero electric field along the sides of the enclosure, and a maximum electric field in the center of the enclosure. Now a PC board is added in the center of the enclosure. The metal planes on the PC board will change the electric field boundary conditions, and a maximum electric field can no longer exist at the enclosure center because the enclosure size has effectively been changed. As other PC boards and subsystems are added, the frequencies where a resonant mode can be supported are reduced to very high frequencies only. In fact, in a normal crowded enclosure, the resonant modes are not likely to be created except at frequencies where the wavelength is in the same range as the separation between subsystem components.

12.3 Shielded Enclosures

For most real world enclosures, the metal thickness is much greater than a skin depth and it is expected that no part of the field will travel through the metal shield. Instead, all leakage will be through apertures and holes in the enclosure. In [12.8] it is discussed how the metal of the shield includes reflection loss and absorption loss for different materials. If it is assumed that the metal is very conductive compared to the air and thick relative to the skin depth, then the effects of the metal shield can be ignored, and the focus placed on the openings and gaskets.

12.3.1 Apertures and Openings

As discussed earlier, when a tangential electric field approaches a metal conductor, the electric field amplitude must go to zero. This creates currents on the surface of the metal conductor. These currents attempt to flow on the enclosure wall surface. If any discontinuity or aperture in the enclosure wall is encountered, the currents must flow around the aperture, as shown previously in Figure 12-2. The need to flow around the aperture increases the path length for the current as also shown in Figure 12-2. The increased path length includes additional inductance and resistance, resulting in an effective voltage across the aperture. If the length of the aperture is less than one-half wavelength for the frequency of interest, the voltage will be at its maximum at the center of the aperture. This voltage is 'seen' on the outside of the aperture (from a source inside the enclosure), and in turn, creates currents on the outside of the enclosure. The external currents cause radiation, so some of the energy from the source inside the enclosure will be radiated externally.

It is clear that when an electric field and the aperture are cross polarized (oriented in the opposite directions) the currents will have the greatest increase in path length, the additional path impedance will be the greatest, and the resulting external fields with be maximized. The longer the aperture, the greater the impedance of the current path, the higher the voltage across the aperture, and therefore the greater the external fields will be. Minimizing the length of the aperture is important to minimizing the amount of 'leakage' through an aperture.

When the direction of the current is the same as the long dimension of the aperture, the current's path is disturbed the least, and the resulting external electric field is minimized. Unfortunately, designers cannot predict or control the polarization of the initial electric field and the resulting current. For effective aperture design, it must be assumed that the current will be flowing in the least desirable direction and design the apertures for this worse-case condition.

Using an array of small holes allows the current to flow in any direction with minimal increased path length, as illustrated in Figure 12-6. The current path is disturbed only minimally. The width of the metal conductor between the hole is important. If too little metal is left between the holes, then the impedance of the current path is increased and the holes will not effectively shield against the fields. A rule-of-thumb for commercial EMC is to keep the width of metal no smaller than 25% of the diameter of the holes to maintain low impedance. Applications where additional shielding is required, such as military applications, will require the width of metal between the holes to be increased.

The shape of the holes is sometimes considered important for the shielding performance of an array of holes. From the previous discussion, it was seen that the shape of the hole is not very important, except if the shape increases the current path length significantly. A round hole provides the least path length through an array with currents from any arbitrary direction. Square holes, diamond shaped holes, etc., result in slightly longer path lengths when the largest diagonal dimension is the same as the round hole diameter for some current directions, but this is not normally significant.

210 / PCB Design For Real-World EMI Control

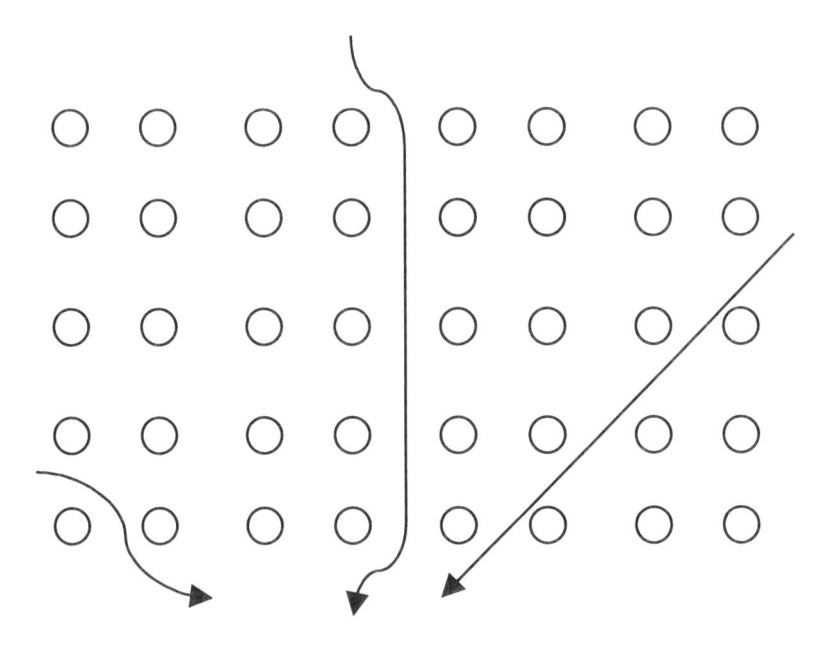

Figure 12-6 Conduction Current on Metal Shield Through an Array of Holes

12.3.2 Gaskets

There is a wide range of gaskets available to help fill in any gaps in the seams of metal covers and make the apparent aperture size smaller. It is important to remember that the goal is to provide a current path across the aperture. If the currents are able to flow across the aperture, there is no increase in the current path length due to the aperture and the resulting voltage across the aperture will be minimized.

The conductivity of the gasket is not the only important parameter for the seam. The surface conductivity of the mating material is very important. Most enclosures are made from a base material, such as steel or aluminum, and a coating is put over the base material to help reduce corrosion. Most often this corrosion coating (e.g. chromate, irridite, etc.) is NOT conductive. Putting a highly conductive gasket over a non-conductive coating does not allow the current to flow across the aperture and defeats the purpose of the gasket. Figure 12-7 illustrates a metal material with a coating and a gasket. When the gasket has roughness, it can cut through the non-conductive coating and make a low impedance contact to the base metal. For this reason, it is best to avoid low cost, smooth cloth-style gasket, since these are not rough enough to cut though any coating material, and may or may not be effective, depending on the thickness of the corrosion protection coating.

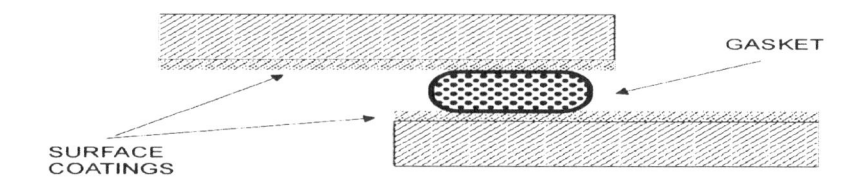

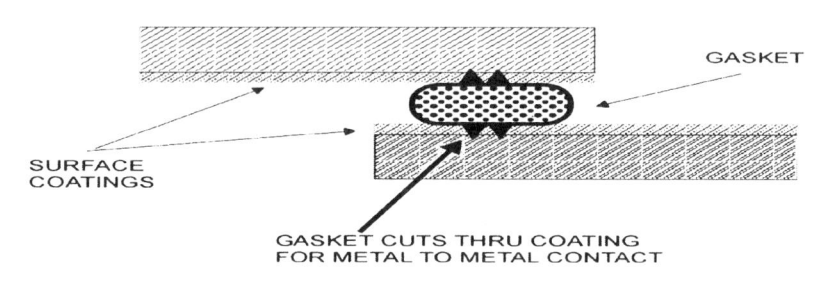

Figure 12-7 Gasket Between Two Metal Panels

Gasket manufacturers will often provide a specification for shielding effectiveness of a particular gasket material. This specification is very misleading. Typically, the shielding effectiveness test is performed using two shielded rooms with a common wall between them. There is a large hole in the common wall. A transmit antenna is placed in one room, and a receive antenna is placed in the second room. A measurement is taken from transmit to receive antennas with the hole open. A gasket material is then placed around the hole and a metal plate covers the hole. The

The location, polarization, and dimensions of the source are not specified in (12.3), since a worst-case envelope independent of these factors is desired. Hence, all of the available power from the source is used. At resonances, this is a reasonable worst-case approximation. The source is assumed to excite modes within the enclosure which will in turn excite leakage through the slots and apertures in the enclosure. Figure 12-9 shows an example of using (12.1) for an enclosure and the agreement with measurements.

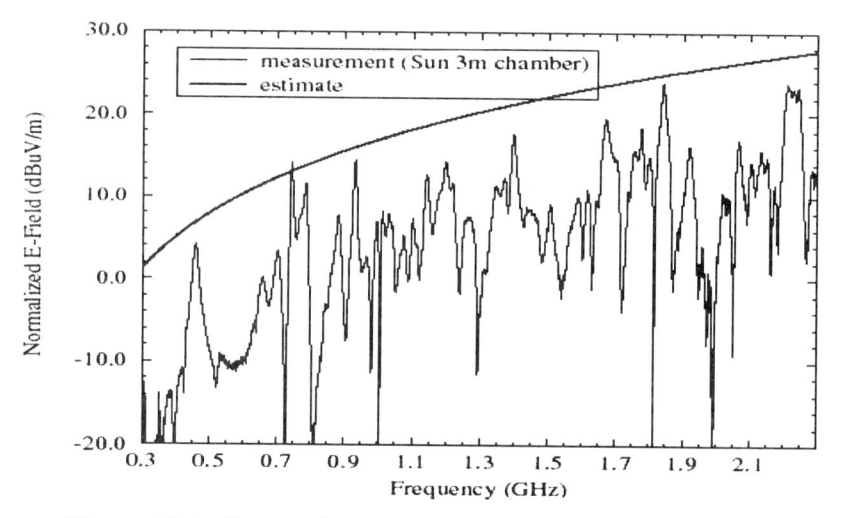

Figure 12-9 Comparison of E-field Worst-Case Estimate and Measured Example

Equation (12.3) can also be given in terms of shielding effectiveness. Assuming a short linear dipole with current I as a noise source in the enclosure, (12.4) is derived using the radiated power from the short dipole as the available power. The resulting shielding effectiveness is then

$$SE = 1.2 \times 10^{12} \frac{\sqrt{\frac{V}{Q}} \ln(1 - 0.66\Delta)}{NL^3 f^{3/2}} \qquad (12.4)$$

12.5 Shielding the PC Board Edge

Previous chapters have discussed the noise between planes on a PC board. A number of decoupling capacitor noise reduction strategies were discussed, but at high frequencies it was very difficult to reduce the noise levels because of the natural inductance of the capacitors and the physical resonances on the board. From an EMI perspective, this noise can travel to the board edge and radiate out from the edge of the board, thus becoming a source for a nearby aperture. One way to reduce the emissions from the board edge is to use a "ground fence" of vias around the outside of the board. Figure 12-10 shows an example of this design layout strategy. An area of ground-reference conductor is routed around the outside of the board on the top and bottom layers. A number of vias are placed around the board edge connecting between the top and bottom layers, creating a partial shield along the PC board edge.

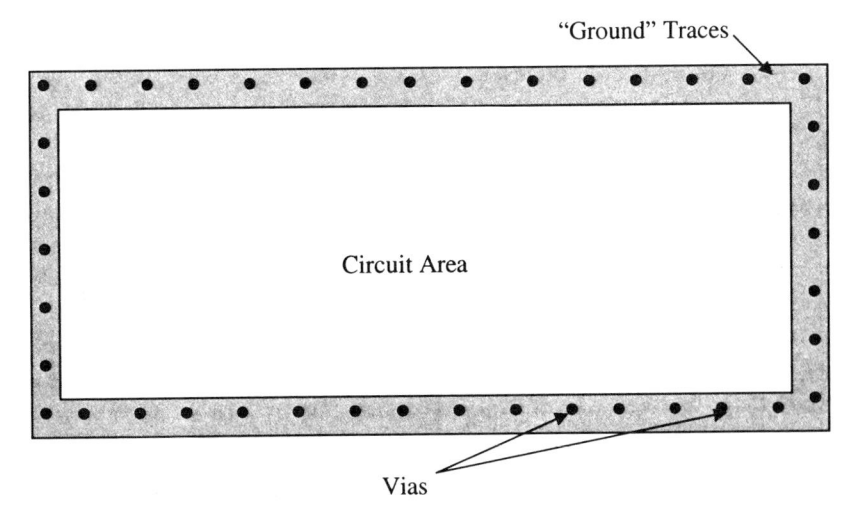

Figure 12-10 "Ground" Stitching Around PCB Edge

Figure 12-11 shows an example of the effectiveness of this strategy for different separations between the vias as a function of frequency. When the vias are placed very close together, this can be an effective design strategy. In the limit, the vias are replaced with solid metal plating around the edge. While solid plating is usually limited to high speed systems, the via fence is often used on lower performance systems. The via spacing is critical and is effective for close via pitch (separation) only.

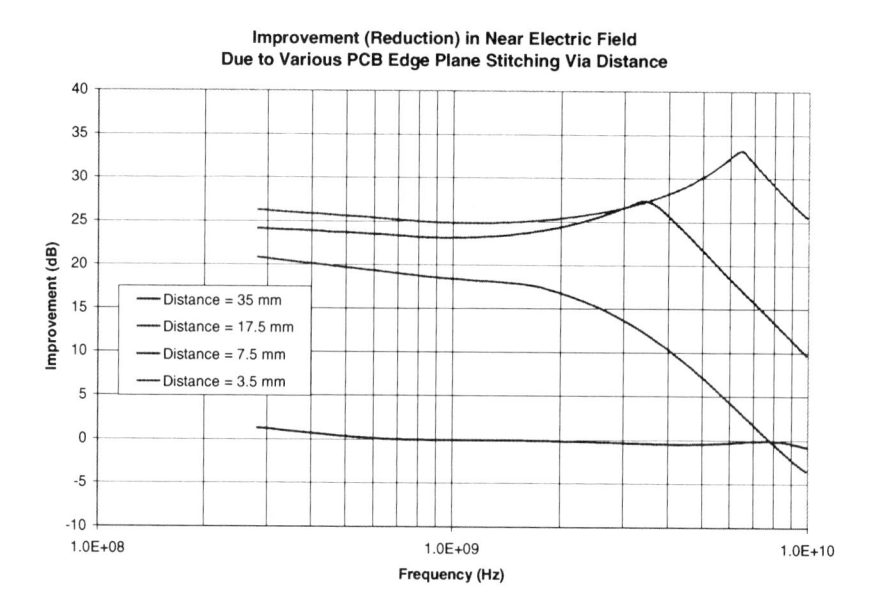

Figure 12-11 Example of Reduction in E-Field from Board Edge for Different Via Spacing

12.6 Cable Shields

Another major leakage point in an otherwise effective enclosure is the shielded cables that are connected to the I/O area of the enclosure. Currents that are not effectively contained within the cable and its connection points to the enclosure will cause external emissions.

In the simplest case of a coax cable, the signal current travels down the center conductor, and returns on the inside of the cable shield. If the cable shield is mated to the enclosure with a 360 degree connection, then all the return current stays inside the cable shield and the enclosure shield. No emissions will occur. If, however, the connection between the cable shield and the enclosure is not a perfect connection, then this impedance, coupled with the current through it, will cause a common-mode voltage to occur between the enclosure shield and the cable shield. This voltage between the cable shield and the enclosure will excite currents across the entire structure, and emissions will occur.

A coax cable is a simple case. Most often, I/O cables are not simple coax cables. A number of wires are typically contained inside the cable shield. These wires may carry high speed signals, differential signals on twisted pairs, or other signals. Intentional or unintentional common mode currents on the internal wires will cause return currents to flow in the inside of the cable shield, and all the concerns described above are again important. In this case, there is a more complex connection between the cable shield and the enclosure chassis. There are a number of different wires involved, and sometimes a variety of different connector types may be used. The connection between the cable shield and the connector backshell and the connection between the connector backshell and the chassis are both critical to the shielding performance. Figure 12-12 illustrates this multi-step connection between the cable shield and the chassis. If either connection point is not a good 360 degree low impedance connection, currents will travel through the impedance of the shield connection and cause a common mode voltage to exist across the poor connection, which in turn causes emissions. It is important to maintain a low impedance connection between each of the parts of the connector system to keep emissions low.

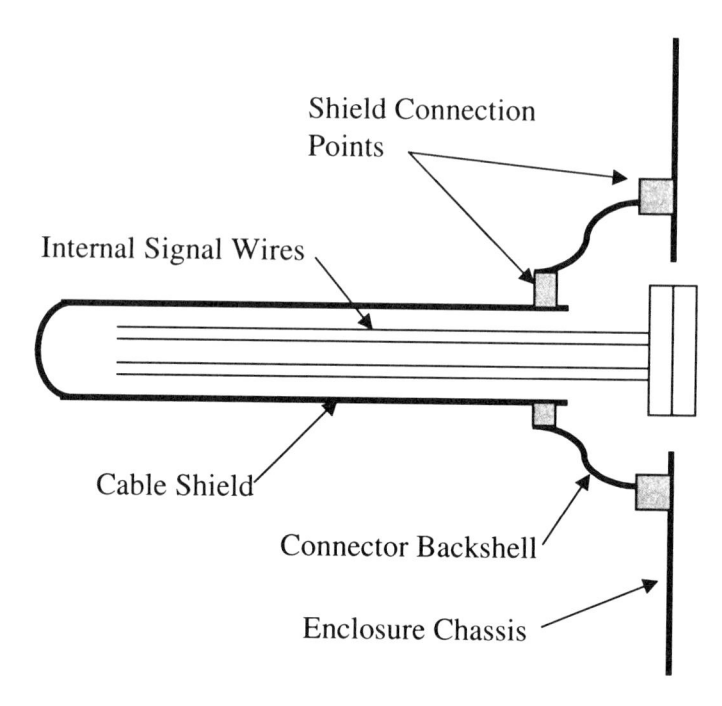

Figure 12-12 Multi-Step Connection Between Cable Shield and Chassis

12.7 Summary

Understanding how enclosure resonances are related to the size of an enclosure and the position of the source is important. It is also important to understand that leakage from seams and other apertures is caused by the current in the metal shield and the added impedance associated with its disturbed path. Minimizing the current path will reduce the effects of the aperture leakage.

Also discussed in this chapter was a technique to predict the leakage from enclosures with typical apertures. While full wave analysis of enclosures will be more exact, the effects of internal resonances are hard to predict since they change as internal wires, boards and other subsystem elements are moved, added, or removed from the enclosure. A simple, worst-case emissions envelope is provided to help determine the over-all effects of the various apertures and openings.

Finally, a brief discussion of cable shield termination to the enclosure chassis shows the need for attention to the cable-to-backshell connection as well as the backshell-to-chassis connection. Using pigtail wires to connect from the cable shield to the chassis or to the "ground" pin on a connector is a poor design choice, and does not maintain an effective shield over the entire system.

References

[12.1] D. M . Pozar, *Microwave Engineering, 2nd Ed.*, John Wiley, New York, 1998.

[12.2] M. Li, J. L. Drewniak, S. Radu, J. Nuebel, T. Hubing, R. DuBroff, and T. Van Doren, "An EMI estimate for shielding enclosure evaluation," *IEEE Trans. Electromagn. Compat.*, vol. 43, pp. 295-304, August 2001.

[12.3] M. P. Robinson, T. M. Benson, C. Christopoulos, J. F. Dawson, M. D. Ganley, A. C. Marvin, S. J. Porter, and D. W. P. Thomas, "Analytical formulation for the shielding effectiveness of enclosures with apertures," *IEEE Trans. Electromagn. Compat.*, vol. 40, pp. 240-248, August 1998.

[12.4] F. Olyslager, E. Laermans, D. De Zutter, S. Criel, R. D. Smedt, N. Lietaert, and A. D. Clercq, "Numerical and experimental study of the shielding effectiveness of a metallic enclosure," *IEEE Trans. Electromagn. Compat.*, vol. 41, pp. 203-213, August 1999.

[12.5] R. Schaumann, and M. E. Van Valkenburg, *Design of Analog Filters*, Oxford University Press, New York, 2001.

[12.6] R. E. Collin, *Field Theory of Guided Waves*, McGraw-Hill, New York, 1960.

[12.7] C. A. Balanis, *Advanced Engineering Electromagnetics*, John Wiley, New York, 1989.

[12.8] C. R. Paul, *Introduction to Electromagnetic Compatibility*, Wiley-Interscience, New York, 1992.

Chapter 13

What To Do If a Product Fails in the EMC Lab

13.1 Introduction

It is a fact of life that products will fail in the EMC test laboratory once in a while. Regardless of whether designers apply all the design strategies explained in this book or not, sometimes events occur that are outside our control. For example, a subsystem is purchased from a vendor and installed in our system. It works fine in other systems, but there is a problem when installed in our system. A shielded enclosure may not be quite as "tight" as planned, etc. The list of things that can go wrong is endless, and to be effective in the testing phase, laboratory engineers must quickly find the source of the problem and propose one or more solutions.

This chapter is intended to describe some strategies useful in helping engineers find the cause of an emissions problem. It is impossible to describe a list of step-by-step tests that will always pinpoint the problem. Each problem is different and the approach must be tailored to the given circumstances. A beginning general test strategy is useful, though.

The general test strategy requires the engineer to be a EMC detective. The first question should always be "Where does this signal come from?" The second question should be, "How does it get out of the shielded box?" Using these two questions as the foundation of your strategy can significantly reduce the amount of time required in the test laboratory. Randomly applying ferrite beads, copper tape, and filter capacitors, on the other hand, is an excellent way to MAXIMIZE your time in the test laboratory and increase cost unnecessarily.

13.2 Where Does the Signal Come From?

Unfortunately, some experienced EMC test engineers tend to ignore this question when working on failing equipment. If you know where the signal is coming from, however, then you can better trace its coupling path to the outside of the enclosure and determine the best solution to the problem.

One of the most likely sources is a clock or data signal. So, what are the clock frequencies and data rates (and their harmonics) in this system? Do any of these frequencies match the problem frequency? Remember, a 100 Mb/s data rate has its fundamental frequency at 50 MHz and possible harmonics every 50 MHz.

Does the spectrum analyzer display give any other information? For example, a common EMC design practice used to lower clock emissions is to use the application of so-called spread spectrum clock signals. This practice causes the clock fundamental frequency and all harmonics to be frequency modulated (typically less than 1% modulation around the fundamental frequency). The resulting display on the spectrum analyzer screen shows a spectrum that has been apparently[1] lowered and "spread" out. Hence the name "spread spectrum". This type of simple frequency modulation has nothing to do with the more traditional spread spectrum communications techniques.

If the spectrum analyzer display of the noise signal shows this spread out emission profile, then the source is most likely from a clock signal that has the spread spectrum option activated. Attention can be devoted to the associated signals and traces on the PC board and all the other signals ignored. The reverse is also true. That is, if the spectrum analyzer display shows a signal that is narrow (not spread out), then any spread spectrum clock can be ignored.

---

[1] The so-called spread spectrum clock signals do not lower the true amplitude of the harmonics. The frequency of the harmonics are changing (through the frequency modulation of the fundamental) and the detector specified in the spectrum analyzer does not have a fast enough charge constant to indicate the true amplitude of the signal.

Another possible indication on the spectrum analyzer display might be neither the narrow single-frequency harmonic, nor the spread spectrum clock spread out display. Emissions from data signals are usually indicated by a somewhat random cluster of signals. Again, understanding the source of the emissions allows the engineer to ignore the obviously unrelated signals on the board.

13.3 How Does the Signal Get Out of the Shielded Enclosure?

Once some idea of where the signal might be originating is determined, one can start looking to see how this signal is getting out of the shielded enclosure. One of the first clues to look for is to see if the signal emission strength varies greatly if various cables are removed and/or moved. Most product shielded enclosures are physically small and not very efficient radiators by themselves. When all the various cables and wires are added to the product, then these cables and wires make the system electrically much larger and a much more effective radiator.

There are really only three ways for signals to get out of the shielded enclosure.
1. Leaks through slots, holes and apertures
2. Conduction through the enclosure shield on cables and unshielded wires
3. Leaks from imperfect mating of shielded cable shields to the enclosure

If a signal exits the shielded enclosure from any of these three ways, moving the cables may change the emission amplitude, depending on how the system is constructed.

13.3.1 Leaks through slots holes and apertures

One common test is to use a magnetic near field probe and run it over the enclosure. While this sometimes helps pinpoint the opening that is causing the emission, it will more often indicate some location other than the one truly causing the problem. In fact, this test will sometimes appear to indicate the solid metal corners or the center of the solid metal panels as the cause of the leak. This false indication is due to the way these near field probes work, that is, they measure the magnetic field due to a current on a surface. Once the signal has leaked out from the opening (wherever the opening is), if the structure is the correct size, a physical size-dependant resonance can be created, and the signal emission enhanced. This resonance will support an RF current on the structure where the peak of the current is controlled by the wavelength of the emission and will have little to do with the location of the actual leak. This behavior is similar to that of a non-center fed dipole antenna. When the frequency is at the half wavelength dipole resonance, the maximum current will be in the center of the antenna, regardless of the location at which the antenna is actually fed.

A better test is to use a contact voltage probe. Figure 13-1 shows a simple example of this probe using a semi-rigid coax cable with the center conductor exposed and the outer conductor extended to the same point as the center conductor. This allows an easy measurement of the voltage (and electric field) across gaps in the shield. A simple version of this probe can easily be created using a standard coax cable quickly when needed.

Figure 13-1 Example RF Contact Probe

To use this probe, the spectrum analyzer is tuned to the frequency of interest, and the probe is physically positioned across each slot and seam of the shielded enclosure. Using this probe, the slots and seams with the highest amount of signal leaking at the frequency of interest can be found very quickly, and temporary copper tape can be applied to determine if a significant reduction is achieved. If the signal cannot be found across any slots and seams, or if using copper tape does not reduce the emission significantly, then the signal is likely to be exiting the enclosure through one of the other means.

If a slot or seam is found to have a high degree of leakage, then gasketing can be applied or the source of the emission can be filtered as appropriate.

13.3.2 Conducted through the shield on cables and wires

Unintentional signals on unshielded cables are a common cause of emissions problems. Disconnecting cables and/or changing their positions can help determine which, if any, cables are responsible for the emission. As in the previous section, it must be kept in mind that detaching a cable might make a significant change in the emission level because the radiating antenna's effectiveness has changed, and not because the source of the emission has been removed!

The simple voltage probe described in the previous section can be used to help isolate which connector is the source of the leakage. The cable can be removed, or the cable can remain in place and the insulation on the cable can be removed to expose the internal conductors. The probe is used to measure the noise voltage between the cable's conductors (or connector pins) and the chassis of the shielded enclosure. CAUTION: to avoid possible damage to the spectrum analyzer check that the intentional signal, or the intentional DC voltage, on the I/O conductor is not too high for the input to the spectrum analyzer.

Depending on how the noise is getting onto the connector pins different results can be seen. Often, the signal is a common-mode signal, and all the pins will have the same level of noise signal energy. This is not always the case, however, and one (or more) conductors could have significantly more noise energy than others. Remember, the emissions are not reading the schematic drawings to see which pins are labeled "ground". Even "ground" pins can have noise energy, depending on how they are connected to the chassis shield. The chassis shield is the ultimate ground-reference for common-mode signals on the external cables. The common-mode noise voltage between the I/O cables and the shielded enclosure chassis drive the cables against the shielded enclosure like two parts of an odd-shaped dipole antenna.

Once the leaking pin(s) are identified, additional filters can be installed on those traces, or filters can be applied at the source of the emission, as appropriate.

13.3.3. Leaks from imperfect mating of shielded cable shields to the enclosure

While it is well known that the cable shield should be a complete 360 degree shield, and that this shield must be connected to the enclosure chassis while maintaining the 360 degree shield coverage, this is seldom practiced except in military applications. Cable shields can be anything from a wrapped foil to a mesh braid, and combinations between them. The quality of the connection from the cable shield to the chassis shield can vary greatly. In the simplest (lowest cost) connection strategy, a drain wire is placed in contact with the foil or braid shield, and then used to connect to the "ground" connector on the system enclosure. More robust solutions will use a metal connector back shell and crimp the cable shield to the connector back shell. The back shell is then mated to a shielded housing on the system enclosure. Regardless of which of the many design strategies is used, the connection between the cable shield and the system enclosure must have low impedance or emissions will occur.

Figure 13-2 shows the currents in a typical cable shield attachment to the system enclosure. The currents can be from intentional signals or unintentional signals. Currents are conducted on the internal conductor(s). The return currents are intended to be only on the inside of the cable shield. If the return currents are constrained to the inside of the cable shield, then no leakage (from the cable) will occur and there will be no emissions problems. If the connection between the cable shield and the chassis of the enclosure does not have very low impedance, however, then a voltage will exist from the flow of this current through the impedance of the shield connection.

What To Do When a Product Fails in the EMC Lab? / 227

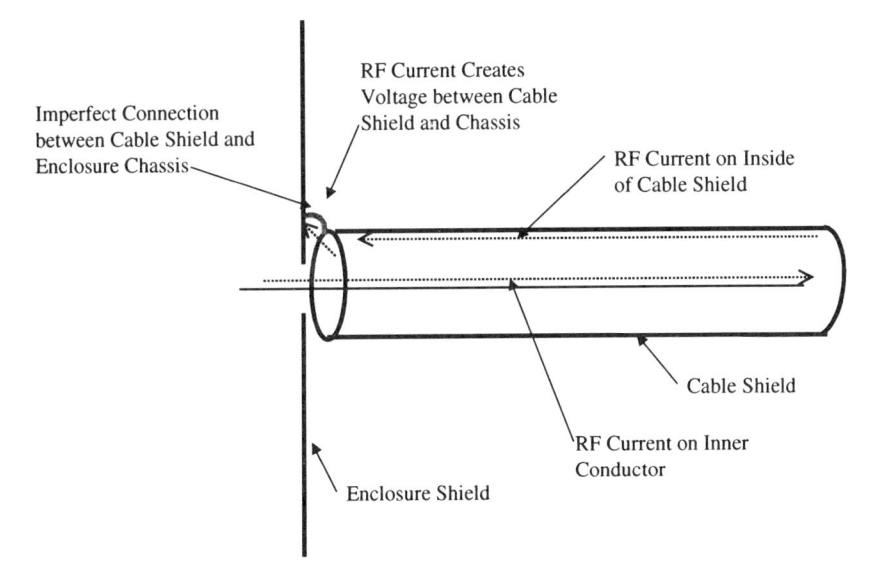

Figure 13-2 Currents on Shielded Cable

The simple voltage probe described in previous sections can be used again to help determine which, if any, of the cable shield connections is the source of the leakage. The probe is used to measure the noise voltage between the cable shield and the enclosure chassis. Once the leaking cable shield connection is identified, the cable shield can be improved, or the signal can be filtered at its source, as appropriate.

13.4 Coupling Mechanism

This chapter has discussed the importance of understanding where the signal comes from and how it gets out of the shielded enclosure, but the coupling mechanism between the source and the ultimate leakage point has not been discussed yet. Often, once a product is failing in the EMC test laboratory, it is too late to do much about the coupling mechanism, but in some circumstances, there are some things can be done to help solve the problem.

13.4.1 Case 1 Clock signal leaking from seam

Assume a clock signal source some place on the PC board and the leakage is from an enclosure seam. If the frequency of the emission is the 7th harmonic or higher of the intentional clock fundamental harmonic, then the best solution is usually to filter the intentional signal. If the fast rise time is truly required for system operation[2], or if the frequency of the problem emission is low enough to be required (first few harmonics), then source filters are not appropriate.

There are many ways for the clock signal to get to the leaking seam. It is most likely that the signal is being radiated inside the enclosure and then leaking from the opening. An internal cable or wire that passes near a seam could be contributing to the leakage. Sometimes moving internal cables will help reduce the emissions (if they are part of the problem). The clock signal could be coupling (through crosstalk) onto another signal conductor that goes into an internal cable and then radiates. Once the coupling mechanism is understood, it can be controlled. There are many different possible coupling paths, and it is not practical to describe them all here.

13.4.2 Case 2 Clock signal leaking from an unshielded cable

The same starting decisions about source filtering as described in the previous section apply again here. However, in this case, the clock signal is conducted out of the enclosure on an unshielded cable. The clock signal is an unintentional signal, since it was never intended to be on that cable. The energy could be coupled from the clock traces onto the I/O traces directly, or through secondary conductors. The energy could be radiated inside the enclosure (as described in the

---

[2] Often, the rise time of clock and data drivers is not required to be as fast as the IC can drive. However, designers will often use fast rise times as a default to make the signal timing analysis easier. This often adds significant cost to the EMC design solutions, since much higher frequencies than necessary must be controlled. As with all designs, a series of trade-offs must be made, and once the additional cost of the EMC solutions is understood, lower signal rise times are often acceptable.

previous section) and then coupled onto the I/O connector and/or traces. Also, as in the previous example, internal cables can act as paths to get the unintentional signal from its source to the I/O connector.

13.5 Summary

There are no step-by-step instructions to help find and overcome EMC problems in the laboratory. However, there is a general strategy to follow to help minimize the time spent solving these problems. Random application of copper tape, ferrite beads and filters is NOT the easiest nor is it the fastest way to find the solution.

Understanding where the signal originates is important. Often, the signal can be controlled at its source, and then it does not matter where the signal leaks from the system enclosure, because source of the signal itself is gone. If the signal cannot be controlled at its source, then designers need to understand how it is leaking from the system enclosure. A simple voltage probe is the most effective analysis tool for this, since it eliminates the possible misdirection from near field probes. This contact voltage probe can be used to help identify leaking enclosure seams, unintentional signals on cables, and poor shielded cable connections. Finally, understanding the coupling mechanism between the source of the signal and the leakage point can help reduce the time to solve the problem.

Appendix A

Introduction to EMI/EMC Computational Modeling

A.1 Introduction

The subject of EMI modeling is beginning to appear in the technical literature with increasing frequency. Most articles identify some new feature or special model that may or may not apply to the general EMI/EMC engineer responsible for product development. Little information is available to the potential user of EMI/EMC modeling tools without requiring reading text books and technical papers containing lots of heavy mathematics and advanced electromagnetic theory.

The current state of the art in EMI/EMC modeling, however, does not require an engineer to have advanced training in electromagnetics or numerical modeling techniques before accurate simulations can be performed and meaningful results obtained. Modeling of EMI/EMC problems can truly help the typical engineer but, like any tool, before modeling can be used effectively, the basics must be understood.

This appendix will serve as a very brief introduction into EMI/EMC modeling techniques. Further introduction into this subject is available in the book "EMI/EMC Computational Modeling Handbook" [A.1]. This appendix will provide a brief introduction to the most popular techniques, and includes some discussion on basic computational modeling philosophy

A.2 Why is EMI/EMC Modeling Important?

The main reason to use EMI/EMC modeling as one of the tools in the EMI/EMC engineer's tool box is to reduce the cost of the product.[1] Without modeling, engineers must rely on handbooks, equations, and graphs, all of which have limited applicability, as well as their own rules-of-thumb, developed through experience. These guidelines are usually based on assumptions that frequently do not exist in the problem at hand. Some guidelines are better, in that they attempt to correct for the inappropriate assumptions, but even these can have accuracy limitations in all but the most carefully controlled circumstances. Proper use of modeling tools allow engineers to use a full-wave electromagnetic solution, rather than one or more simplifications, to predict the effect in the specific product of concern.

Given the limitations that these guidelines have in most real world problems, engineers are faced with either a conservative or a non-conservative design. The conservative design will insure that the product will meet the appropriate regulatory limits the first time. This can be assured only by over-design of the EMI/EMC features. This over-design will usually meet the appropriate limits, but extra cost is added to the product. The non-conservative design will take some reasonable chances to reduce the amount of EMI/EMC features required. Depending on the engineer's experience and training, the product may or may not meet the regulatory limit. If the product doesn't meet the limit, a panic redesign is required, most often resulting in product ship delays and extra cost due to the band-aid nature of such "fixes".

Another realistic benefit of the use of EMI/EMC modeling is credibility. Often the product design team consists of a number of different engineering disciplines: electrical, mechanical, thermal, and EMI/EMC. Computer Aided Design (CAD) simulation tools are commonly used in other engineering disciplines. These tools provide significant credibility to the engineer's claim for whatever design features they recommend to be included for a successful EMI/EMC

---

[1] The cost of the product can be measured in both the development costs and the time-to-market costs.

design. These features, such as larger air vent openings, are often in direct conflict with the EMI/EMC engineer's design direction; however, since the EMI/EMC engineer has no simulation to rely on, their recommendations are often ignored. EMI/EMC modeling tools can provide the design team with reliable numerical results, taking the guesswork out of the design, and providing the EMI/EMC engineer with the credibility to get their design recommendations seriously considered by the team.

A.3 EMI/EMC Modeling: State of the Art

Current EMI/EMC modeling tools cannot do everything. That is, they cannot take the complete mechanical and electrical CAD files, compute overnight, and provide the engineer with a green/red light for pass/fail for the regulatory standard desired. The EMI/EMC engineer is needed to reduce the overall product into a set of problems that can be realistically modeled. The engineer must decide where the risks are in the product design, and analyze those areas.

This means that the EMI/EMC engineer must remain an integral part of the EMI/EMC design process. Modeling will not replace the EMI/EMC engineer. Modeling is only one of the tools that EMI/EMC engineers have at their disposal. The knowledge and experience that the EMI/EMC engineer uses during the design process is needed to determine which area of the design needs further analysis and modeling.

Often, the problem to be analyzed will require a multi-stage model. The results of one model's simulation will provide the input to the next stage model. This allows the model to be optimized for each particular portion of the problem, and the results combined. Thus, much larger overall problems can be analyzed than by using a brute force approach, in which the entire problem is modeled at once. Again, the EMI/EMC engineer needs to understand the problem and the modeling techniques well enough to know where to break it into individual simulations.

234 / PCB Design For Real-World EMI Control

A.4 Tool Box Approach

No single modeling technique will be the most efficient and accurate for every possible model needed. Unfortunately, many commercial packages specialize in only one technique, and try to force every problem into a particular solution technique. The EMI/EMC engineer has a wide variety of problems to solve, requiring an equally wide set of tools. The "right tool for the right job" approach applies to EMI/EMC engineering as much as it does to building a house or a radio. You would not use a putty knife to cut lumber, or a soldering iron to tighten screws, so why use an inappropriate modeling technique?

A wide range of automated EMI/EMC tools are available to the engineer. Automated tools include design rule checkers that check Printed Circuit Board (PCB) layout against a set of pre-determined design rules; quasi-static simulators, which are useful for inductance/capacitance/resistance parameter extraction when the component is much smaller than a wavelength; quick calculators using closed-form equations calculated by computer for simple applications; full-wave numerical simulation techniques as described in this book; and expert-system tools, which provide design advice based on a predetermined set of conditions. It is clear that these different automated tools are applied to different EMI/EMC problems, and at different times in the design process. This book will focus on the full-wave numerical modeling and simulation techniques, and how to apply these techniques to real-world EMI/EMC problems.

Different modeling techniques are suited to different problems. For example, the Method of Moments (MoM) technique is perfectly suited for a long wire simulation, since it only determines the currents on conductors, such as metal surfaces and wires, and it is independent of the volume of free space around the wires. However, the Finite Element Method (FEM) and the Finite-Difference Time-Domain (FDTD) method are not well suited to model long wire simulations, such as a computer with an external cable, since they require a volume of space to be modeled around the wire, and this volume must be large enough to have the computational domain boundary in the far field. Thus, using the FEM or the FDTD techniques for these applications results in a computationally inefficient model. On the other hand, there are problems for which the MoM is not a suitable choice, therefore, a set of tools that contain different modeling techniques is a great asset to the EMI/EMC engineer.

A.5 Brief Description of EMI Modeling Techniques

There are a variety of electromagnetic modeling techniques. Which is the '"best" technique is cause for a significant amount of debate, and often becomes a matter of which school the developer attended, and which technique his or her professor specialized in. Many of the techniques are specialized for certain configurations, and require cumbersome tailoring when used for each problem. Some techniques are not particularly generic, and require in-depth knowledge of electromagnetics and the modeling technique. Still others are useful only for far-field problems, such as determining a radar cross section of a piece of military equipment. None of these specialized far-field techniques will be discussed here, since they have little use for the typical EMI/EMC engineer's problems.

Three techniques are typically used for EMI/EMC modeling problems: the FDTD technique, the MoM, and the FEM technique. Each technique will be briefly described here and then in greater detail in Chapters 3 through 5, so the EMI/EMC engineer can better understand how and when to use them.

A.5.1 Finite-Difference Time-Domain

The FDTD technique is a volume-based solution to Maxwell's differential equations. Maxwell's equations are converted to central difference equations, and solved directly in the time domain. The entire volume of space surrounding the object to be modeled must be gridded, usually into square or rectangular grids. Each grid must have a size that is small compared to the shortest wavelength of interest, and have its location identified as metal, air, or whatever material desired. Figure A.1 shows an example of such a grid for a two- dimensional case. Once the grid parameters are established, the electric and magnetic fields are determined throughout the grid at a particular time. Time is advanced one time step, and the fields are determined again. Thus, the electric and magnetic fields are determined at each time step based on the previous values of the electric and magnetic fields.

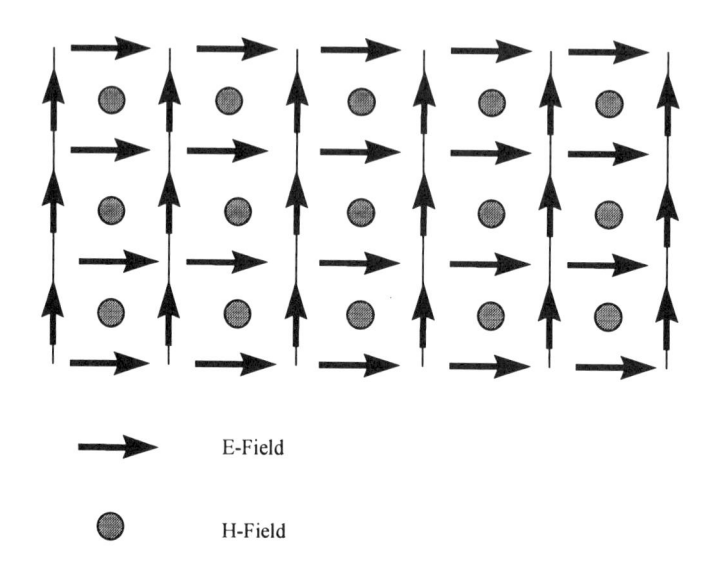

→ E-Field

● H-Field

Figure A.1 Two-Dimensional FDTD Grid

Once the fields have propagated throughout the meshed domain, the FDTD simulation is complete, and the broadband frequency response of the model is determined by performing a Fourier transform of the time-domain results at the specified monitor points. Since the FDTD method provides a time-domain solution, a wide band frequency-domain result is available from a single simulation.

Since the FDTD technique is a volume-based solution,[2] the edges of the grid must be specially controlled to provide the proper radiation response. The edges are modeled with an Absorbing Boundary Condition (ABC). There are a number of different ABCs, mostly named after their inventors. In nearly all cases, the ABC must be electrically remote from the source and all radiation sources of the model, so that the far-field assumption of the ABC holds true, and the ABC is reasonably accurate. Typically, a good ABC for the FDTD technique will provide an effective reflection of less than –60 dB.

---

[2] The entire volume of the computational domain must be gridded.

Naturally, since the size of the gridded computational area is determined from the size of the model itself, some effort is needed to keep the model small. The solution time increases as the size of the computational area (number of grid points) increases. The FDTD technique is well suited to models containing enclosed volumes with metal, dielectric, and air. The FDTD technique is not well suited to modeling wires or other long, thin structures, as the computational area overhead increases very rapidly with this type of structure.

A.5.2 Method of Moments

The MoM is a surface current technique.[3] The structure to be modeled is converted into a series of metal plates and wires.[4] Figure A.2 shows an example of a shielded box converted to a wire grid with a long attached wire. Once the structure is defined, the wires are broken into wire segments (short compared to a wavelength) and the plates are divided into patches (small compared to a wavelength). From this structure, a set of linear equations is created. The solution to this set of linear equations finds the RF currents on each wire segment and surface patch. Once the RF current is known for each segment and patch, the electric field at any point in space can be determined by solving for each segment/patch and performing the vector summation.

When using the MoM, the currents on all conductors are determined, and the remaining space is assumed to be air. This facilitates the efficiency of the MoM in solving problems with long thin structures, such as external wires and cables. Since the MoM finds the currents on the conductors, it models metals and air very efficiently. However, dielectric and other materials are difficult to model using the MoM with standard computer codes.

The MoM is a frequency-domain solution technique. Therefore, if the solution is needed at more than one frequency, the simulation must be run for each frequency. This is often required, since the

---

[3] Only the surface currents are determined, and the entire volume is not gridded.
[4] Often, a solid structure is converted into a wire frame model, eliminating the metal plates completely.

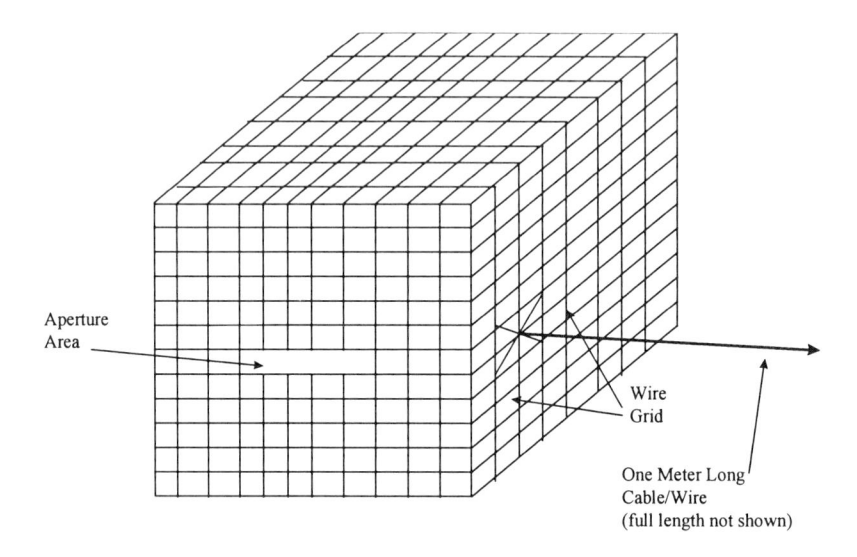

Figure A.2 MoM Wire Mesh Model of Shielded Enclosure with 1 meter Long Cable Attached

source signals within the typical computer have fast rise times, and therefore wide harmonic content.

A.5.3 Finite Element Method

The FEM is another volume-based solution technique. The solution space is split into small elements, usually triangular or tetrahedral shaped, referred to as the finite element mesh. The field in each element is approximated by low-order polynomials with unknown coefficients. These approximation functions are substituted into a variational expression derived from Maxwell's equations, and the resulting system of equations is solved to determine the coefficients. Once these coefficients are calculated, the fields are known approximately within each element.

As in the above techniques, the smaller the elements, the more accurate the final solution. As the element size become small, the number of unknowns in the problem increase rapidly, thus increasing the solution time.

The FEM is a volume-based solution technique; therefore, it must have a boundary condition at the boundary of the computational space. Typically, the FEM boundaries must be electrically distant away from the structure being analyzed, and must be spherical or cylindrical in shape. This restriction results in a heavy overhead burden for FEM users, since the number of unknowns is increased dramatically in comparison to other computational techniques.

A.6 Other Uses for Electromagnetics Modeling

Although this book will focus mostly on EMI/EMC modeling, and converting those types of problems into realistic models, there are many other uses for modeling. Antenna design, radar cross section and microwave circuit analysis are only a few. These types of problems tend to have software focused especially on those problems; however, the techniques used for EMI/EMC modeling may be easily applied to these other specialized problems. In general, the most effective EMI/EMC modeling engineers will take an electromagnetic view of the overall problem, breaking it into source and receive for analysis.

A.7 Summary

EMI/EMC problems are here to stay, and becoming more complex as personal communication devices proliferate and computer speeds continue to rise. Every electronic product on the market today and planned for tomorrow requires EMI/EMC considerations. Those engineers who insist on performing product design using previous methods only will quickly find themselves with design projects that are too expensive, either because of over-design or because of repeated design cycles before regulatory compliance is reached. Although not every design project, nor every EMI/EMC design feature must be modeled, modeling/simulation can be a very useful tool to engineers. Experience has shown that once the initial hesitation to use something new is overcome, engineers find ways to use the modeling tools that they had never previously imagined.

References

[A.1]   B. Archambeault, C. Brench, O. Ramahi, "EMI/EMC Computational Modeling Handbook". 2nd Edition, Kluwer Academic Publishers, 2001

Index

aperture
   current around, 202
apertures, 208
ASIC/IC
   power current demand, 125
   shoot-through current, 126
asymmetrical stripline, 48

buried capacitance, 144

cable shields, 216
capacitors
   decoupling, 122
   lossy, 146
   non-ideal, 164
   stitching, 72
   typical values, 122
clock signals
   harmonic content, 14
coax cable, 217
common-mode currents, 4
common-mode filters, 168
component placement, 195

computational modeling, 231
computer aided design, 232
connector pin assignments, 82
contact probe, 224
coupling mechanisms, 10, 227
critical signals, 86
   through vias, 100
crosstalk, 113, 114, 115, 116
   cascade, 114
   controlling, 115
   multi-level, 114
crosstalk analysis, 185

Daughter Cards, 80
DDRAM, 179
decoupling, 121 - 149
decoupling capacitor
   purpose, 147
   effectiveness, 130
   source decoupling, 141
   global (distributed), 136
   selection of value, 140

decoupling noise
  from ASIC/ICs, 124
  source, 124
differential signal lines, 181

edge rate, 89
electric field coupling, 10
electromagnetic induction, 25
EMI Sources, 4
  intentional signals, 85
emissions
  from external wire, 112
emissions sources, 21
  shielded products, 21
  unshielded products, 22

Faraday's Law, 26
FDTD, 235
FEM, 235
ferrite beads
  non-ideal, 166
filter configurations
  low pass, 154
  three-component, 161
  two-component, 155
filter design, 151
  concepts, 151
filters
  common-mode, 168
Finite-Difference Time-Domain, 235
Finite Element Method, 238

gaskets, 210
ground, 6, 43 - 66
  chassis reference, 50
  earth safety reference, 55
  I/O Area, 64
  power reference, 49
  signal reference, 48
  symbols, 45
ground stragety
  multi-point, 55
  single-point, 55

harmonic content, 14
  edge rate, 89
  non-squarewave, 18
  squarewave, 87
  trapezoidal pulse, 88
heatsink, 10
  grounding, 58
inductance, 5, 25 - 41
intentional current
  spectrum, 172
intentional signal, 23, 86
interrupted Return Path, 99
I/O filter, 52
  configurations, 153
I/O Filter Design, 151
isolation, 195

layout
  PCB, 187
lossy capacitors, 146

magic, 9, 20
magnetic field coupling, 12
Method of Moments, 237
microstrip line, 48
MoM, 235
mutual inductance, 27
   arbitary loops, 28
   coaxial oriented, 28

non-ideal
   capacitors, 164
   ferrite beads, 166
   Zero-Ohm resistors, 168

partial inductance, 36, 53
PCB
   Stack-up, 187
PCB Layout, 187
potential emissions
   common-mode (intentional signal), 96
   common-mode unintentional signals, 106
potential sources
   loop-mode, 93
   unintentional signals, 105
power current requirements from ICs, 125
probe
   contact, 224

reference planes
   changing, 76

impedance across split, 111
split, 71, 108
return current, 69, 78, 79, 82, 83
routing strategies, 81
resonance, 19
   cavity, 20
   TE mode, 205

SDRAM, 178
self inductance, 29
   equilateral triangular loop, 31
   isosceles triangular loop, 32
   pair of wires in free space, 33
   two flat co-planar traces, 36
   two flat traces, 35
   wire over a metal plane, 34
shielded Cables, 54
shielding, 6, 199
shielding effectiveness
   predicting, 213
shielded enclosures
   resonance, 202
signal integrity, 90, 92
signal integrity tools, 171
signal spectra, 14
simulation techniques
   Finite Element Method, 238

Method of Moments, 237
Finite-Difference Time-
Domain, 235
skin-effect, 54
squarewave
harmonic content, 87
spread spectrum, 222
stack-up
four-layer, 192
many layer boards, 188
one and two layer, 193
six-layer, 191
PCB, 187
stitching capacitors, 72
symmetrical stripline, 48

TEM, 69
TEMPEST, 2
termination resistor, 173
testing, 221
two-component filters
reference connection, 157

unintentional signals, 106
unshielded cables, 51